SpringerBriefs in Applied Sciences and Technology

SpringerBriefs present concise summaries of cutting-edge research and practical applications across a wide spectrum of fields. Featuring compact volumes of 50 to 125 pages, the series covers a range of content from professional to academic.

Typical publications can be:

- A timely report of state-of-the art methods
- An introduction to or a manual for the application of mathematical or computer techniques
- A bridge between new research results, as published in journal articles
- A snapshot of a hot or emerging topic
- An in-depth case study
- A presentation of core concepts that students must understand in order to make independent contributions

SpringerBriefs are characterized by fast, global electronic dissemination, standard publishing contracts, standardized manuscript preparation and formatting guidelines, and expedited production schedules.

On the one hand, **SpringerBriefs in Applied Sciences and Technology** are devoted to the publication of fundamentals and applications within the different classical engineering disciplines as well as in interdisciplinary fields that recently emerged between these areas. On the other hand, as the boundary separating fundamental research and applied technology is more and more dissolving, this series is particularly open to trans-disciplinary topics between fundamental science and engineering.

Indexed by EI-Compendex, SCOPUS and Springerlink.

More information about this series at http://www.springer.com/series/8884

Jagadish · Sumit Bhowmik

Manufacturing and Processing of Natural Filler Based Polymer Composites

 Springer

Jagadish
Department of Mechanical Engineering
National Institute of Technology Raipur
Raipur, Chhattisgarh, India

Sumit Bhowmik
Department of Mechanical Engineering
National Institute of Technology Silchar
Silchar, Assam, India

ISSN 2191-530X   ISSN 2191-5318   (electronic)
SpringerBriefs in Applied Sciences and Technology
ISBN 978-3-030-65361-3   ISBN 978-3-030-65362-0   (eBook)
https://doi.org/10.1007/978-3-030-65362-0

This Springer imprint is published by the registered company Springer Nature Switzerland AG
The registered company address is: Gewerbestrasse 11, 6330 Cham, Switzerland

Preface

At present, usage is increased in many industrial and domestic applications of polymer composite materials filled with natural fibers, also known as natural fiber-based reinforced polymers (NFPs), because of the rising environmental concerns and the need for more versatile manufacturing materials. In the same way, the NFRP composites are reinforced by various natural fibers such as oil palm, sugar palm, bagasse, coir, banana stem, hemp, jute, sisal, kenaf, roselle, rice husk, betel nuts husk, bamboo, cane, and coir Among the other polymer composites, NFRP composite gives particular points of interest, for example, minimal effort, dimensional security, low thickness, adaptability, satisfactory explicit quality, sustainable asset, and they are appropriate to be used in different applications, for example, car bodyboards, tooling, furniture businesses, building ventures, bundling materials, and so on. Besides, physical, chemical, and mechanical characterization, manufacturing, and processing/machining of NFRP composites are very limited. In fact, manufacturing methods [namely injection molding, compression molding, resin transfer molding, hand layup method, vacuum-assisted resin transfer molding, etc.] for NFRP composites are similar to those synthetic fiber composites but some additional accessories may be required. Contrarily, NFRP composites undergo various secondary operations and also know machining operations (such as drilling, trimming, cutting, and milling) before going to the actual use and to get the final product. Thus, a study on the machining or processing of NFRP composites is essential before actual use.

This book gives the important points of the material manufacturing, machining experimentation, modeling, and optimization of conventional machining and non-conventional machining methods in various machining scenarios on NFRP composites carried out by using the authors. It discusses the research results of secondary processing such as processing of different natural fiber (namely bamboo, hemp, banana, and jute) based polymer composites specifically using CNC drilling, milling and AWJM processes under different machining environments like dry machining, green machining/manufacturing. Besides, various MCDM techniques (single and integrated) like entropy-MCRA, MOORA, MOOSRA, and AHP-TOPSIS in order to get optimal trust force, process time, material removal rate, process energy, surface quality properties. This book is also intended to address some environmental issues such as reduction of wastages, use of different cutting fluids issues, and energy consumption. related to the machining or processing.

The present book contains five chapters starting with an introduction to NFRP composites followed by manufacturing methods, and their processing/machining issues are discussed in Chap. 1. Chapter 2 presents the manufacturing and machining of short bamboo fiber-based polymer (SBFRP) composites and optimization of the CNC penetrating procedure in dry machining combined with trial research for ideal qualities thrust force, material removal rate, and point angle coordinated entropy-multi-models positioning investigation. Chapter 3 discussed the hemp fiber-based polymer (HFRP) composites followed by parametric analysis, ANOVA, empirical models, and optimization to get optimal values for CNC drilling process in tap water using fuzzy-MOORA. Chapter 4 reports on banana fiber-based polymer (BFRP) composites related to its manufacturing and processing coupled with statistical analysis (parametric analysis, ANOVA, and empirical models) and optimization CNC milling process in dry machining parameters using AHP-TOPSIS. Chapter 5 discussed the manufacturing and processing of jute fiber-based reinforced polymer (JFRP) composites followed by application of integrated method, i.e., fuzzy-MOOSRA to progress in the performance of AWJM process along with ANOVA and empirical models.

The authors acknowledge to Dr. Shailesh Vaidya, National Institute of Technology Raipur, India, for providing the necessary resources and other facilities during the research work and also thanks to Mr. Sathish Kumar Adapa, Assistant Professor, Department of Mechanical Engineering, AITAM, AP, and Mr. Shubam B. Patil for his necessary help during the work.

Authors hope that the work reported in this book would help and inspire the engineers, researchers, and experts working in a similar field.

We really acknowledge Springer for this probability and their expert support.

Raipur, India   Jagadish
Silchar, India   Sumit Bhowmik

Contents

Nomenclature

$A_1, A_2 \dots A_m$	Number of alternatives
B_{ij}	Normalized evaluation matrix for beneficial criterion
$C_1, C_2 \dots C_n$	Number of criteria
D_{ij}	Decision matrix of ith alternatives on jth criteria
d_j	Degree of divergence
N_{ij}	Normalized performance values of ith parameters on jth criteria
NB_{ij}	Normalized evaluation matrix for non-beneficial criterion
$P_{11}, P_{12}, \dots P_{1n}$	Criterion and the major weights of two criteria
S_i^+ and S_i^-	Euclidean distances positive (+ve) and negative (–ve) ideal solution
U_{ij}	Weighted normalized decision matrix of ith alternatives on jth criteria
U^+ and U^-	Matrix of positive (+ve) and negative (–ve) ideal solution
x	Pairwise comparison matrix of ith alternatives on jth criteria
y_j	Assessment values of ith parameters w.r.t. jth criterion
$Y_{11}, Y_{12}, \dots Y_{mn}$	Response values of m alternatives on n criterion
ζ_j	Weights of jth criteria
ω_{ij}	Weighted decision matrix of ith alternatives on jth criteria
α and β	Summation of weighted response values of jth criteria

Notations

C_T	Cutting time in s
F_W	Final weights in g
I	Current in Amps
I_W	Initial weights in g
V	Voltage in volts
X	Load applied in N
ρ	Density of fiber in g/mm^3

Acronyms

AHP	Analytic hierarchy process
ANOVA	Analysis of variance
AWJM	Abrasive water jet machining
BFRP	Banana fiber-based polymer
CM	Conventional machining
CNC	Computer numerical control
EDM	Electrical discharge machine
FRP	Fiber-reinforced polymer
GFRP	Glass fiber-reinforced polymer
HFRP	Hemp fiber-reinforced polymer
JFRP	Jute fiber-reinforced polymer
MCDM	Multi-criteria decision making
MCRA	Multi-criteria ranking analysis
MOORA	Multi-objective optimization ratio analysis
MOOSRA	Multi-objective optimization on the basis of simple ratio analysis
NCFR	Newspaper cellulose fiber-reinforced
NCM	Non-conventional machining
NF	Natural fiber
NFRP	Natural fiber-based reinforced polymer
PLA	Poly lactic acid
PVC	Poly(vinyl chloride)
RTM	Resin transfer molding
SBF	Short bamboo fibers
SBFRP	Short bamboo fiber-reinforced polymer
TOPSIS	Technique for order of preference by similarity to ideal solution
USM	Ultrasonic machining
VARTM	Vacuum-assisted resin transfer molding
WFRP	Wood fiber-reinforced polymer

Symbols

CC	Closeness coefficient
CI	Closeness index
CS	Cutting speed
LB	Larger-the-better
MRR	Material removal rate
OA	Orthogonal array
PA	Point angle
PE	Process energy
PT	Process time
SB	Smaller-the-better
SR	Surface roughness
SS	Spindle speed
SS	Spindle speed
TF	Thrust force

Chapter 1
Introduction

Abstract This chapter introduces to manufacturing and processing or machining of natural fiber-based reinforced polymers (NFRP) composites. It starts with an introduction and classification of NFRP composites followed by manufacturing methods, namely injection molding; compression molding; resin transfer molding; hand layup method; and vacuum-assisted resin transfer molding and processing or machining of NFRP composites. Based on the previous studies, the statistics as regards to the manufacturing and processing or machining of NFRP composites are discussed. The chapter ends with detailed summary with possible scope of future work on manufacturing of natural fibers (namely bamboo, hemp, banana, and jute) based polymer composites and their detailed processing/machining using both conventional and non-conventional machining processes. The chapter also highlighted about green machining aspects and optimization of machining processes.

Keywords NFRP · Machining · Natural fibers (NF) · CM · NCM

1.1 Natural Fiber-Reinforced Polymer Composites

At present, usage is increased in many industrial and domestic applications of polymer composite materials filled with natural fibers, also known as natural fiber-based reinforced polymers (NFRP), because of the rising environmental concerns and the need for more versatile manufacturing materials [1–3]. In the same way, the NFRP composites are reinforced by various natural fibers like jute, bamboo, hemp, coir, banana, sisal, kenaf, rice husk, cane, coir, etc. [4–7]. The main advantages of NFRP composites are low cost, renewable, plenty, lightweight, and less abrasive, and hence, they are appropriate to use in components of semi- or non-structural engineering.

© The Author(s), under exclusive license to Springer Nature Switzerland AG 2021   1
Jagadish and S. Bhowmik, *Manufacturing and Processing of Natural Filler Based Polymer Composites*, SpringerBriefs in Applied Sciences and Technology,
https://doi.org/10.1007/978-3-030-65362-0_1

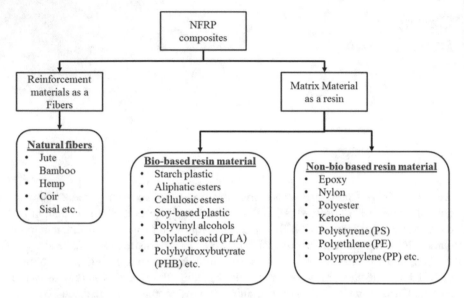

Fig. 1.1 Composition of NFRP composites

Natural fiber-based reinforced polymer composites consist of two major components (Fig. 1.1): one being the matrix and other being reinforcement. The main function of matrix material is to bind the reinforcement materials with it while the function of reinforcement is to provide the strength to the composites and helps in the transfer of the load to the matrix material. The most common matrix material used in NFRP composites is based on biodegradable (starch plastic, aliphatic esters, cellulosic esters, soy-based plastic, etc.) and non-biodegradable resins like epoxy, nylon, polyester, ketone, etc. Mostly, all kinds of natural fiber (Fig. 1.2) like sugar palm, bagasse, coir, banana stems, hemp, jute, coir, etc. are available in nature, and they are used as reinforcement material in the fabrication of NFRP composites, because natural fibers are abundant in nature; low-cost processing; and also possess different properties compared to the synthetic fibers.

In addition to the composition of NFRP composites, their poor stability during processing or manufacturing has often been difficult. These composites are mainly manufactured in primary and secondary production methods. In fact, the most popular primary manufacturing method, namely hand layup, injection molding, resin and vacuum-assisted resin transfer molding, gives the final component a near-net form. In addition, these near-net types also require some secondary machining operations to be specific turning, milling, and drilling to meet the actual requirement of the final components for assembly. Processing or machining is also particularly important to enable assembly of the parts. It requires slot cutting with various shapes and a high superior finish and higher output at minimal machining costs and maximum

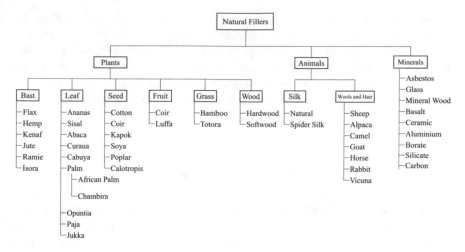

Fig. 1.2 Natural fibers used in NFRP composites

machining efficiency. Furthermore, before actual uses for finished products, these composites are subject to comprehensive secondary manufacturing or machining operations.

The essential characteristics of natural fiber-based reinforced polymer composites include durability, biodegradability, good mechanical properties, lower costs, water and rot, dimensional stability, and easy processing capability for several applications. The major applications of these composites are transportation (i.e., railway, aerospace, automobiles, etc.) as replacement of different plastic-based matrix materials like polypropylene, polyethylene, building and construction industries (like particleboard, ceiling paneling, etc.), packaging and consumer products, etc. [8]. According to the literature survey [9], the average cost of glass fiber is between $1.30 and $2.00/kg while natural fiber prices are between $0.22 and $1.10/kg [10]. Without a secondary process, all the abovementioned applications are impossible, i.e., machining or processing during the manufacturing of final products. Thus, machining or processing of these composites is highly essential.

Furthermore, machining or processing of natural fiber-based reinforced polymer composites using conventional machining and some of the non-conventional machining (NCM) processes bit different and difficult due to their superior properties such as inhomogeneous and anisotropic. Due to their relatively low heat conductivity, conventional machining causes excessive heat at the cutting zone, and there are difficulties in the dissipation of the heat. As a result, it decreases the tool life, increases the surface roughness, and decreases the dimensional sensitivity of work materials. To resolve, many conventional machining processes use cutting fluid during the heating issues in the machining process. But, the use of different cutting fluids in machining results in various forms of environmental issues or toxic substances in the waste form that contribute to severe occupational health and environmental problems. Hence, machining under a green manufacturing environment is highly essential. Therefore, in this book authors have employed green manufacturing techniques like the use of tap water as a dielectric fluid, dry machining, and green machining process, i.e., abrasive water jet machining (AWJM) process for machining or processing of NFRP composites. Mainly four types of natural fiber (bamboo, jute, hemp, and banana) based polymer composites have been manufactured and machined using conventional machining and non-conventional machining processes under green manufacturing conditions.

1.2   Natural Fibers Used in NFRP Composites

As environmental problems are growing and materials are required for more flexible production, material engineers and researchers are studying composites with natural fillers. To overcome most issue, i.e., environmental, engineers are using various types of natural fibers (NFs). Fibers are hair-like materials that can be divided into filaments, threads, ropes for various uses. NFs are divided into three main categories based on plant, animal, and mineral origin as shown in Fig. 1.2. The fiber under plants is further classified as bast, leaf, seeds, fruit, grass, and wood fibers. Some of the few examples of plan-based fibers are flax, hemp, kenaf (bast); bananas, abaca, curaua (leaf); coir, cotton, soya (seeds); luffa, pineapple (fruits), etc. Animal-based fibers are classified as wood and hair and silk [11, 12]. Few examples are sheep, alpaca, camel (wools and hair), and natural silk and spider silk (silk). The most commonly used mineral fibers are glass, asbestos, silicate, carbon, etc.

Apart from the classification, natural fibers possess some unique mechanical properties when compared to the glass fiber-based and other man-made-based composites, NFs-based composites provide lesser mechanical strength. But due to their specific properties like low density, strength, and stiffness are a bit comparable to the man-made composites. The mechanical characteristics are based on the cellular material and the microfibrillar angle of the natural fiber with respect to the natural fiber-reinforced polymer (NFRP) composites. In the case of the natural fiber, the Young's modulus increases with decreases in the diameter of the fiber. From the literature, the study suggested that natural fiber with high cellulose content and low microfibrillar angles viable for polymer composites. The mechanical properties of the most used NFs are shown in Table 1.1.

1.3   Manufacturing Methods of NFRP Composites

In this section, manufacturing methods reported in the literature related to the development of manufacturing of natural fiber's based polymer composites are discussed. Based on the research of Wang et al. [6] on manufacturing methods of NFRP composites, mainly five major manufacturing methods, namely injection molding; compression molding; resin transfer molding; hand layup method; and vacuum-assisted resin

Table 1.1 Details of fiber properties [13, 14]

Fibers	Tensile strength (MPa)	Young modulus (GPa)
Natural fiber		
Cotton	285–595	5.5–12.5
Coir	170–175	
Flax	340–345	27–30
Hemp	670–690	
Jute	390–775	55–60
Kenaf	380–935	35–55
Pineapple	170–175	60–65
Ramie	410–940	61.4–128
Sisal	515–640	9.4–28
Spider silk	875–975	13–15
Man-made fiber		
Aramid	3000–3200	65–70
Carbon	4000	230–240
E-glass	2000–3500	75

transfer molding, are being used in both academic and in industry. In fact, among all the conventional techniques, most of the researchers suggested that hand layup method is convenient compared to others, because this method offers some of the unique advantages such as no complicated equipment is required, simple steps, less investment, and greater flexibility. Hence, in this book, authors have used a hand layup method for the manufacturing of different NFRP composites. The brief descriptions of each of the manufacturing methods are explored as follows.

1.3.1 Injection Molding Method

Injection molding processes are recommended for thermoplastic-based processing. This method is one of the most commonly used plastic production methods. This process can also be used in the manufacture of composites made from fiber-reinforced polymers. But the fiber requirement should be in the form of short or micron size. The schematic diagram of the typical injection molding system is shown in Fig. 1.3.

An injection molding with the fitted extruder screw is nurtured into a heated barrel by means of a hopper. Shearing action of the reciprocating screw and heated barrel caused to melt the mixer in the barrel (Fig. 1.3). Then, a molten mixer is injected into the mold using pressure load to get the required fiber-reinforced polymer (FRP) composite object. After that, FRP composite object is ejected/removed from the mold via ejecting pins with some suitable pressure force. Many studies have been conducted in this direction, including Huda et al. [15], in comparison with the

Fig. 1.3 Schematic diagram of an injection molding machine

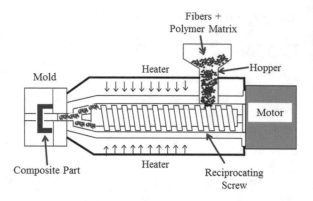

glass fiber-reinforced polymer (GFRP) and the recycled newspaper cellulose fiber-reinforced polymers (NCFRs) in the injection molding process for the manufacturing of the composites with injection molding procedures. The mechanical properties like tensile and flexural properties of NCFR-PLA were studied and found greater than the GFRP composites. Further, a study on the dispersion of fiber into the polymer matrix was done via morphological analysis through SEM found that uniform distribution of both chopped glass and recycled paper fibers into the PLA matrix. Azaman et al. [16] have done the simulation of wood fiber-reinforced polymer (WFRP) composite behaviors in the injection molding process using Autodesk MoldFlow 2011 software. Wood fiber reinforced polymer composite flow was simulated in the mold behavior and the correct properties such as residual stress, shrinkage, and thin-walled components' warpage values were calculated.

1.3.2 Compression Molding Method

The compression molding is a high-pressure, high volume plastic molding process, suitable for molding composite materials with high strength. Many production lines have chosen to manufacture compression molds because it offers many advantages, namely short cycle times, dimensional accuracy, low-cost production, lesser impact resistance, uniform shrinking, and density.

Compression molding process consists of two stages: preheating and pressurizing. In the preheating stage, first, both the polymer matrix and the fiber molten binding are preheated and placed in the corresponding mold cavity (Fig. 1.4a). Secondly, the squeezing of preheated molten mixer (polymer + fiber) in the mold cavity is done by applying a suitable pressure and heat (Fig. 1.4b). Finally, the mold is removed, and then the composite component is removed subsequently. The pressure is usually applied in this phase from 14 to 20 MPa at 150–190 °C. Parameters such as composite density, strength, and orientation on the fibers affecting the process are careful in the preparation of the composite, and the random application of extreme pressure can lead to a fiber fracture. Fibers should be mounted easily inside the mold to

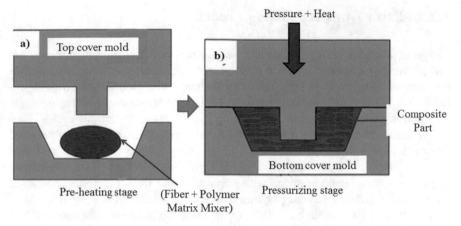

Fig. 1.4 Compression molding process: **a** Preheating stage, **b** pressurizing stage

solve these problems. It is also possible to propose a natural fiber premix with a polymer compound. This process may result in the use of long fibers to generate bio-composites with a higher fracture in volume.

In the literature, NFRP composites manufactured by compression molding have been studied by various researchers such as Wirawan et al. [17] developed sugarcane bagasse reinforced poly(vinyl chloride) PVC using compression molding method followed by mechanical properties characterization. First, blend of sugarcane bagasse fiber + PVC mixer were preheated and placed in the mold and then applied suitable pressure and heat to squeeze the preheated molten mixer into the mold cavity to form sugarcane bagasse-based polymer composite. The intermediate materials were placed into the mold before applying the pressure and heat to prevent the damage on the part. Usually, a temperature of 170 °C and 14 MPa pressure was used during the preparation of the composite part. Later, different mechanical characterization like tensile, flexural, modulus, etc. was studied on the formed composite part and found that chemically treated sugarcane bagasse-based polymer composite shows better results compared to the untreated one. In another, the research of El-Shekeil et al. [18] has shown the applicability of the compression molding method in the preparation of kenaf fiber with PVC composites and its characterization. The fiber content of the composite variety: 20, 30, and 40% were prepaid with process parameters of 140 °C, 11 min, and 40 rpm respectively for temperature, time, and speed. Later, different mechanical characterization like tensile properties, flexural properties, impact strength, etc. was studied on the formed composite and found that kenaf fiber with PVC composites shows better results. Additionally, thermal stability was studied and found better stability.

1.3.3 Resin Transfer Molding Process

The resin transfer molding (RTM) is a well-known manufacturing method and has low pressure (vacuum-assisted), low-speed molding process for many well-known bio-composite applications [19]. The complete steps wise process of RTM is shown in Fig. 1.5. Resin transfer molding process consists of four stages, i.e., molding stage, injection stage, curing stage, and demolding stage. First, in the molding stage, fiber preform was developed and placed in the mold cavity. Here, a long fiber obtained from nature was first to cut into short fibers using a knife or scissors, and then fiber preforms kept in a mold cavity (Fig. 1.5a). Second, short fibers also known as preforms were bound or mixed with polymer matrix (resin) material and closed the mold (Fig. 1.5b). Here, the polymer matrix (resin) material is being transferred into the mold cavity through a variety of equipment like pressure or vacuum. The two matching molds are compressed tightly to prevent any resin leakage during the injection.

A former treatment stage to ensure that the resin is cured completely is recommended in this process (Fig. 1.5c). Within the resin, colors and other improvements within resin usually are applied to minimize costs, to prevent problems like decreasing, to increase flame resistance, and to increase mechanic properties, for polyester and Vinyl Ester resins in which the load-bearing is not important. At last, demolding, i.e., mold, is open, and the composite part is ejected from the mold (Fig. 1.5d). One of the researches of Salim et al. [20] is shown the application of RTM in the manufacturing of kenaf fiber-reinforced epoxy polymer composites followed by mechanical properties determination and optimization. Here, the stitching density of nonwoven kenaf fiber mat was optimized and found that stitched kenaf mat composite yields better mechanical properties compared to unstitched ones.

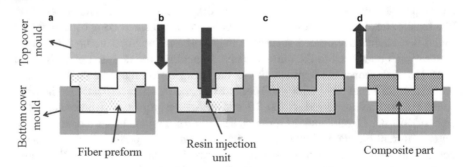

Fig. 1.5 Resin transfer molding process: **a** Molding stage, **b** injection stage, **c** curing stage, **d** demolding stage

1.3.4 Hand Layup Process

The hand layup method can be seen in Fig. 1.6, which is the most commonly used manufacturing procedure for open mold composites. This process consists of five steps, i.e., mold preparation, coating, pouring, rolling, and demolding. First, the mold of the required size is being prepaid depending upon the application (Fig. 1.6a). Second, to ease the removal of the finished composite component, a thin layer of an anti-adhesive coat is applied on the mold surface (Fig. 1.6b). This thin layer will act as a replicant and prevents the fiber mixer to attach to the mold bottom surface. Then, resin (polymer matrix) with fiber mixer is poured into the open mold and simultaneously applied the suitable pressure on the molten mixer using a roller (Fig. 1.6d). Here, a roller is used for the proper distribution of mixer in the mold and also ensuring the desired thickness of the part is obtained. Then, composite is cured at room temperature or an oven for 24–30 h. At last, demolding is being done, and the composite part is taken from the mold (Fig. 1.6e). Necessary care must be taken during the demolding stage to avoid damages in the part. Hence, high skilled labor is recommended to perform the fabrication task in this process.

In the literature, many investigators used this method for preparation of NFRP composites such as glass-sugar palm fiber-reinforced unsaturated polyester composites by Misri et al. [21], wood dust reinforced polymer composites and pineapple leaf fiber-reinforced polymer composites by Jagadish et al. [22, 23], coir, and wood fiber-based polymer composites by Kumar et al. [24, 25]. In all these studies, authors have performed many physical, chemical, mechanical characterizations.

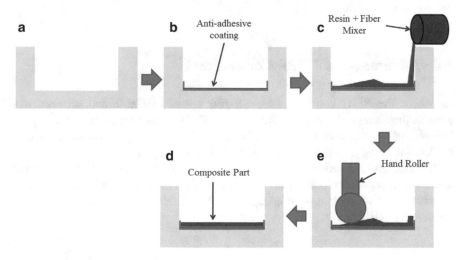

Fig. 1.6 Hand layup process: **a** Molding stage, **b** coating stage, **c** pouring stage, **d** rolling stage, **e** demolding stage

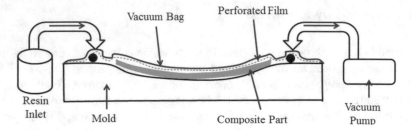

Fig. 1.7 VARTM process

1.3.5 VARTM Process

Similar to the RTM, vacuum-assisted resin transfer molding (VARTM) is a recent development in the composite manufacturing industry. The fibers are placed between a vacuum bag and a resin container, and a perforation pipe is positioned in this process. As shown in Fig. 1.7, the vacuum force causes the resin to be sucked over the fibers through perforated tubes/film to stabilize the laminate structure. Within the compound structure, excess air is not permitted in this method and is common for the development of large objects such as boat hulls and wind turbine blades [19]. In this method, to improve the strength of the composite, surface treatment of natural fiber is recommended. The research of Ishikawa et al. [19] was alkali treated flax, hemp with epoxy acrylate resin-based polymer composites manufactured by VARTM showed an increment of 19.7% strength compared to the unalkali treated polymer composites.

1.4 Advantages of NFRP Composites

Generally, the NFRP composites have higher specific strength and stiffness than glass as they have low specific weight. Also, NFRP is renewable, and they require less energy for production. They also emit less CO_2 during production and reduce production costs. The processing is easy and non-toxic, and there are no wear of tools and no skin irritation. They have high electrical resistance and good thermal and acoustic insulation properties. They are biodegradables and recyclables. They are cost-effective, have low weights, available abundantly, and most importantly environmentally friendly.

1.5 Limitation of NFRP Composites

However, there are some limitations to NFRP composites. They can be processed and used in a restricted temperature. They have lower impact strength. They are easily influenced by the weather, e.g., more moisture causes swelling of the fibers. Hence, they have low durability. They have poor fire resistance and also have poor fiber/matrix adhesion if the fibers are not properly processed. Moreover, the irregularity in the harvest and agricultural results leads to price fluctuation.

1.6 Applications of NFRP Composites

The NFRP composites have been used to replace the current higher-weight materials and for various applications. The solid composites were created with the hybrid composites of oil palm/coir fiber. The low-cost bio-fiber hybrid composites have been designed for structural cellular plates. The various applications of NFRP composite materials are listed with respective specific fields like aerospace, automobile, construction, sports, etc. The various applications of natural fibers are illustrated in Fig. 1.8. The graphical representation shows the highest application of these composites in the consumer and textiles industry [22] followed by automotive, building and construction, packaging, etc. Hence, there is a huge scope of these composites in several sectors that need to be explored. Some of the important application products wise in the different sector are given detail as follows.

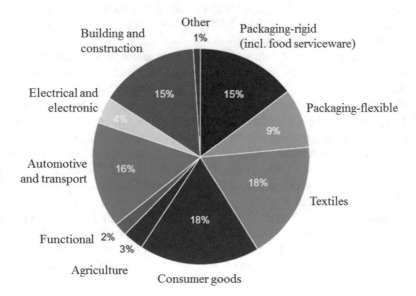

Fig. 1.8 Applications of NFRP composites in various fields

One of the growing applications of NFRP composites is making panels for partition and false ceiling, partition boards, walls, floors, windows, door frames, roof tiles, etc. in the building and construction industry [26]. Another important application is in the aerospace sector, fabricating tails, wings, propellers, helicopter fan blades [27, 28]. The use of NFRP is also growing in the packaging industries as well, especially for thermal insulating and gas barrier packaging. Some other uses include furniture: chair, table, shower, bath units, electrical appliances, pipes, automobile, railway coach interior, boat, gears, ice skating boards, bicycle frames, baseball bats, tennis racket, fork, helmet, post-boxes, etc.

1.7   Processing/Machining of NFRP Composites

In many engineering structures, interest in natural fiber-based reinforced polymer composites increases rapidly. The wide variety, the lack of expertise of engineers, limited knowledge of databases for processing, setting criteria, and manufacturing complexity as barriers to the widespread use of composites. In the literature, many works are available on machining of fiber-reinforced polymer composites [29, 30], while review on machining of NFRP composites is scarce.

Machining of composites NFRP in particular created many problems, because of its mechanical anisotropies and heterogeneous, which are not usually found in the machining of the traditional metallic/monolithic material. Hence, machining of NFRP composites finds difficult and costly attempt compared to metal [30] and should be explored. In fact, NFRP composites are made to near-net shape, yet these composites need auxiliary or secondary manufacturing processes, for example, machining to meet the final product that incorporates in assembly and dimensional prerequisites.

The machining of NFRP composites requires a thorough understanding of the method of cutting and its optimal design to increase the quality and precision of the manufacturing [31]. Composites machining is mainly affected by various parameters like type of natural fibers used, their properties, kind of matrix material used and their properties, the orientation of fiber in the matrix, etc. Hence, the selection of appropriate machining processes as well as a proper understanding of cutting mechanisms is essential. Machining defects found most commonly in the manufacturing of NFRP composites are delaminations, debond, matrix imperfection, resin-starved area, resin-rich area, voids, cracks, fiber pull-out, thermal damages, etc. [32–35]. In addition to the defects, machining of NFRP composites depends on machining performance parameters. The most commonly used machining performance parameters are cutting force, cutting power, tool wear, and tool life. There are other factors such as tool material, tool geometry, cutting conditions, and work materials that influence the quality of the machining. Besides, the other quality parameters such as surface roughness, torque, thrust force, process time, process energy, delamination, and MRR have not given much attention. Furthermore, the machining of NFRP composites generates various environmental harmful components or toxic substances in the waste form that lead to severe occupational health and environmental problems. Thus, in this book these all the above aspects are considered for the study.

1.8 Summary and Recommendation

This chapter presented a quick overview of the introduction to the manufacturing and processing of NFRP composites primarily based on the literature study. It has been observed that investigations on the primary operation of different natural fiber with respect to physical, chemical, mechanical, and thermal characterization were done by part researchers. However, the researches on secondary operation, i.e., processing/machining of NFRP composites, are yet to be explored. Various manufacturing techniques were employed for the manufacturing of fiber-reinforced polymer composites. The compression molding and conventional hand layup methods are found more convenient, productive, and economical for producing NFRP composites compared to others. Except for these parameters like tool material, tool geometry, cutting conditions, and work materials, other parameters quality parameters such as surface roughness, torque, thrust force; process time, process energy, delamination, and MRR have not given much attention during the processing/machining of NFRP composites. Furthermore, environmental aspects like wastage generation, process energy, and process time consumption during the processing/machining of NFRP composites have not been explored. The above observations encourage further research on the manufacturing and processing of NFRP composites. The subsequent chapter presents the details of the original experimental work performed by the authors on the manufacturing of natural fibers (to be specific short bamboo, hemp, banana, and jute) based polymer composites and their detailed processing/machining using both conventional and non-conventional machining processes. In addition, optimization of machining processes is discussed using individual and integrated multi-criteria decision making (MCDM) methods.

References

1. M.A. Fuqua, S. Huo, C.A. Ulven, Natural fiber-reinforced composites. Polym. Rev. **52**, 259–320 (2012)
2. M. Zimniewska, J. Myalski, M. Koziol, J. Mankowski, E. Bogacz, Natural fiber textile structures suitable for composite materials. J. Nat. Fibers **9**, 229–239 (2012)
3. M.J. John, K.T. Varughese, S. Thomas, Green composites from natural fibers and natural rubber: effect of fiber ratio on mechanical and swelling characteristics. J. Nat. Fibers **5**, 47–60 (2008)
4. M.P. Westman, S.G. Laddha, L.S. Fifield, T.A. Kafentzis, K.L. Simmons, *Natural Fiber Composites: A Review*, Report No. PNNL-19220. Report prepared for US Department of Energy, p. 1–10 (2010)

5. M. Ho, H. Wang, J. Lee, C. Ho, K. Lau, Critical factors on manufacturing processes of natural fibre composites. Compos. Part B Eng. **43**, 3549–3562 (2012)
6. J. Vera-Sorroche, The effect of melt viscosity on thermal efficiency for single screw extrusion of HDPE. Chem. Eng. Res. Des. **92**(11), 2404–2412 (2014)
7. A.K. Mohanty, M. Misra, G. Hinrichsen, Biofibres, biodegradable polymers and biocomposites: an overview. Macromol. Mater. Eng. **276**(1), 1–24 (2000)
8. J. Zajac, Z. Hutyrová, I. Orlovský, Investigation of surface roughness after turning of one kind of the bio-material with thermoplastic matrix and natural fibers. Adv. Mater. Res. **941–944**, 275–279 (2014). https://doi.org/10.4028/www.scientific.net/AMR
9. D. Chandramohan, K. Marimuthu, Drilling of natural fiber particle reinforced polymer composite material. Int. J. Adv. Eng. Res. Stud. **1**, 134–145 (2011)
10. R. Gachter, H. Muller, *Plastics Additives*, 3rd edn (Hanser Publishers, Hanser Verlag, Munich, 1990), p. 5–970
11. F.P. La-Mantia, M. Morreale, Green composites: a brief review. Compos. Part A Appl. Sci. Manuf. **42**, 579–588 (2011)
12. S. Taj, M.A. Munawar, S.U. Khan, Natural fibre-reinforced polymer composites. Proc. Pakistan Acad. Sci. **44**(2), 129–144 (2007)
13. D.N. Saheb, J.P. Jog, Natural fibre polymer composites: a review. Adv. Polym. Technol. **18**(4), 351–363 (1999)
14. M.S. Huda, L.T. Drzal, A.K. Mohanty, M. Misra, Chopped glass and recycled newspaper as reinforcement fibers in injection molded poly(lactic acid) (PLA) composites: comparative study. Compos. Sci. Technol. **66**, 1813–1824 (2013)
15. A. Memon, A. Nakai, Mechanical properties of jute spun yarn/PLA tubular braided composite by pultrusion molding. Energy Procedia **34**, 818–829 (2013)
16. M.D. Azaman, S.M. Sapuan, S. Sulaiman, E.S. Zainudin, A. Khalina, Shrinkages and warpage in the processability of wood-filled polypropylene composite thin-walled parts formed by injection molding. Mater. Des. **52**, 1018–1026 (2013)
17. R. Wirawan, S.M. Sapuan, R. Yunus, A. Khalina, Properties of sugarcane bagasse/poly (vinyl chloride) composites after various treatments. J. Compos. Mater. **45**, 1667–1674 (2011)
18. Y.A. El-Shekeil, S.M. Sapuan, K. Abdan, E.S. Zainudin, Influence of fiber content on themechanical and thermal properties of kenaf fiber reinforced thermoplastic polyurethane composites. Mater. Des. **40**, 299–303 (2012)
19. H. Ishikawa, H. Takagi, A.N. Nakagaito, M. Yasuzawa, H. Genta, H. Saito, Effect of surface treatments on the mechanical properties of natural fiber textile composites made by VaRTM method. Compos. Interface **21**, 329–336 (2014)
20. M.S. Salim, Z.A.M. Ishak, S.A. Hamid, Effect of stitching density on non-woven fibre matto-wards mechanical properties of kenaf reinforced epoxy composites produced by resin transfer moulding (RTM). Key Eng. Mater. **471–472**, 987–992 (2011)
21. S. Misri, Z. Leman, S.M. Sapuan, A small boat from woven glass-sugar palm fibre reinforced unsaturated polyester composite, in *Engineering Composites: Properties and Applications*, ed. by S.M. Sapuan (UPM Press, Serdang, 2014), pp. 297–312
22. Rajakumaran M. Jagadish, A. Ray, Investigation on mechanical properties of pineapple leaf based short fiber reinforced polymer composite from selected Indian (north eastern part) cultivars. J. Thermoplast. Compos. Mater. **33**(3), 324–342 (2018)
23. Bhowmik S. Jagadish, A. Ray, Prediction of surface roughness quality of green abrasive water jet machining: a soft computing approach. J. Intell. Manuf. **30**, 2965–2979 (2015)
24. R. Kumar, K. Kumar, S. Bhowmik, Optimization of mechanical properties of epoxy based wood dust reinforced green composite using Taguchi method, in *International Conference on Advances in Manufacturing and Materials Engineering*; NIT Surathkal, Published in Proc Mater Sci, vol. 5, p. 688–696 (2014)
25. R. Kumar, K. Kumar, S. Bhowmik, Assessment and response of treated Cocos nucifera reinforced toughened epoxy composite towards fracture and viscoelastic properties. J. Polym. Environ. **26**(6), 2522–2535 (2017)

26. P.J. Davim, L.R. Silva, A. Festas, A.M. Abrão, Machinability study on precision turning of PA66 polyamide with and without glass fiber reinforcing. Mater. Des. **30**, 228–234 (2009)
27. M.R.M. Jamir, M.S.A. Majid, A. Khasri, Natural lightweight hybrid composites for aircraft structural applications, *Sustainable Composites for Aerospace Applications* (Woodhead Publishing, Sawston, UK, Cambridge, UK, 2018), pp. 155–170
28. A.T. Marques, Fibrous materials reinforced composites production techniques, *Fibrous and Composite Materials for Civil Engineering Applications* (Woodhead Publishing, Sawston, UK Cambridge, UK, 2011), pp. 191–215
29. J.P. Davim, P. Reis, Machinability study on composite (polyetheretherketone reinforced with 30% glass fibre–PEEK GF 30) using polycrystalline diamond (PCD) and cemented carbide (K20) tools. Int. J. Adv. Manuf. Technol. **23**, 412–418 (2004)
30. J.P. Davim, F. Mata, V.N. Gaitonde, S.R. Karnik, Machinability evaluation in unreinforced and reinforced PEEK composites using response surface models. J. Thermoplast. Compos. Mater. **23**, 5–18 (2010)
31. F.D. Ning, W.L. Cong, Z.J. Pei, C. Treadwell, Rotary ultrasonic machining of CFRP: a comparison with grinding. Ultrasonics **66**, 125–132 (2016)
32. Y. Karpat, O. Bahtiyar, B. Deger, Milling force modelling of multidirectional carbon fiber reinforced polymer laminates. Proc. CIRP **1**, 460–465 (2012)
33. M. Zampaloni, F. Pourboghrat, S.A. Yankovich, Kenaf natural fiber—a discussion on manufacturing problems and solutions. Compos. Part A Appl. Sci. Manuf. **38**(6), 1569–1580 (2007)
34. D. Babu, K.S. Babu, B.U.M. Gowd, Drilling uni-directional fiber-reinforced plastics manufactured by hand lay-up influence of fibers. J. Mater. Sci. Technol. **1**, 1–10 (2012)
35. L. Durão, J. Tavares, V. de Albuquerque, J. Marques, O. Andrade, Drilling damage in composite material. Mater. (Basel) **7**, 3802–3819 (2014)

Chapter 2
Manufacturing and Processing of Short Bamboo Fiber-Based Polymer Composite

Abstract At present, the application of natural fiber-based polymer composites are widely increased in various industries due to their greater specific strength, eco-friendly, biodegradable, and low cost. Recently, bamboo fiber is being used as reinforcement in the fabrication of polymer composites for the applications especially automotive, aerospace, and consumer industries. Drilling under presence of dielectric fluid is essential processing operations that are frequently carried out on the composites to facilitate the assembly process in manufacturing the engineering products. However, the presence of dielectric fluid and drilling tool alone produces an ample of hazardous components in the form of wastes, health, and environmental issues during the processing. Hence, in the present chapter, an attempt is made on manufacturing and processing of short bamboo fiber-based reinforced polymer (SBFRP) composites under dry processing environment. This study aimed at manufacturing of composites using compression molding technique and evaluation of processing parameters (spindle speed, cutting speed, and point angle of drill bit) on thrust force, MRR, and process time in processing. ANOVA and multi-regression analysis are employed to study the statistical significance. The optimal setting for CNC drilling is found through fuzzy-MOOSRA method. The experimental results indicated that the CNC drilling with dry cutting environment is capable of processing the SBFRP composites to produce high quality parts with excellent productivity and less process time, thrust force.

Keywords SBFRP composite · CNC drilling process · Dry machining · ANOVA · Regression analysis · Optimization

2.1 Introduction

Bamboo is the most widely used natural fiber plant across the world for a variety of day-to-day applications. Bamboo (commonly known as *Bambusoideae*) is an ancient farming crop in India, at an approximated area of 10 million hectares and almost all cover 13% of the total forest area. In India the total production of bamboo per year is around 5 million tonnes. Bamboo-based fillers are cultivated in large quantities in the

Jagadish and S. Bhowmik, *Manufacturing and Processing of Natural Filler Based Polymer Composites*, SpringerBriefs in Applied Sciences and Technology,
https://doi.org/10.1007/978-3-030-65362-0_2

entire region of India such as Assam, Tripura, Meghalaya, Manipur, Chhattisgarh, and West Bengal. More than 1000 species of bamboo plants are available, converted into two groups *herbaceous* and *woody* [1]. The bamboo plant consists of several parts like stem or culm (it contains sulcus, internode, node, branch, and sheath), steam base, and steam petiole. The bamboo stem or the culm is most beneficial part of the bamboo plant. The pull-out fiber from the stems can be applied in a variety of applications such as cloth, automobile, furniture, and sports industry as an alternative to synthetic fiber [2]. Moreover, bamboo fillers/fibers provide unique advantages such as inexpensive, good establishment, low thickness, more flexible, more secure to the operator, greater specific strength, eco-friendly, and biodegradable, and this indicates the utilization of industry in distinct medium load applications such as automotive body panels, assembly, chair, table, construction, and packaging materials [3–6]. Bamboo can be found generally green, yellow, pink, and black or even combinations in color, and the plant is grown approximately 25 m height with a diameter varying from 10 to 200 mm. The chemical composition of bamboo consists of cellulose ($\approx$80–90%), hemicelluloses (40–50%), lignin ($\approx$5–12%), ash ($\approx$1.1%), and other extractives. In the present chapter, bamboo fillers/fibers are selected as reinforcement material due to a high degree of crystalline and cellulose content [3, 4, 7]. Meanwhile, some extent of work has been done on the bamboo fiber-based composites with their chemical and mechanical properties. The chemical treatment on bamboo fibers with sodium hydroxide solution (NaOH) or combination of NaOH and sodium sulfite Na_2SO_3 solution is done. The result shows that moisture resistance and mechanical properties are improved [8]. Mechanical properties of bamboo fiber-reinforced epoxy composites, chemically treated with NaOH (6% w/v). The result shows that the mechanical properties are significantly increased with increasing the % fiber content and length of the fiber [9]. Yu et al. [10] physical and mechanical properties of bamboo fiber content and their effect on the properties is studied. It is observed that increase in fiber content increases the mechanical properties of the composites. Huda and Reddy [11] evaluated the mechanical properties of bamboo strip-based polymer composite. Poly-propane is used as matrix materials and compared the results with glass fiber-based polymer composites. Manalo et al. [12] alkali treated bamboo fiber base polymer composite and their mechanical properties determination is examined. Namely, NaOH treatment (48% by weight) is done and properties of the composites are evaluated. Work recommended that processing of bamboo fiber enhances the mechanical properties compared to the untreated fibers. Similar work also carried out by Huang and Young [13] on extraction of mechanical, hygrothermal and interface power of continuously properties of treated bamboo fiber and found greater improvement in mechanical properties compared to the untreated.

Besides, physical, chemical, and mechanical characterization, natural fiber or bamboo fiber-based reinforced polymer composite needs some secondary manufacturing operations like turning, milling, and drilling to meet the actual requirement of the final components for assembly. Hence, processing/machining of NFRP (like short bamboo fillers-based) composites is highly essential in manufacturing the quality engineering products. Babu et al. [14] studied the machining behavior of hemp fiber-based polymer composite using Taguchi technique and ANOVA and suggested

optimal selection of cutting parameters for better machining. Naveen et al. [15] investigated the drilling parameters like speed and feed on the damage factor of hemp fiber
and glass fiber-based composite. Three fraction of fiber volume (i.e., 10, 20, and 30%)
fiber content of the composite are prepared. Result reported that higher feed fibers
are difficult to cut, and some of the uncut fiber was present during the machining.
Similar study is also done by Yallew et al. [16] in which hole damage in drilling is
studied using stereo microscope considering the thrust force and torque parameters
for jute-based polymer composites. Work reported that test results were obtained as
very close to the acceptable level. Wang et al. [17] studied the traditional drilling and
hybrid ultrasonic drilling (UAD) on hemp fiber composite laminate. Result reveals
that machining in UAD is better than traditional drilling process. Chandramohan
and Marimuthu [18] examined the machining performance of natural fiber polymer
composite materials using drilling process. Work reported that thrust strength and the
torque increase with the drill diameter and feed rate and for larger drills and higher
feed speed, a greater thrust force occurs during machining. Babu et al. [19] had
done the comparative study of machining properties evaluation between glass fiber
and natural fiber-based polymer composites using traditional milling process. Input
conditions like feed and speed and output conditions such as delamination factor and
surface ruggedness were considered. Result shows that at higher speed and lower
feed rate gave minimum delamination and good surface roughness for natural fiber-
based polymer composites. Jayabala et al. [20] performed machinability study of coir
reinforced polymer composites. Regression equations developed for making input
(drill diameter, spindle speed, and feed rate) and output (thrust force, torque, and tool
wear) in drilling analysis. Work established the optimal conditions (6 mm drill diameter, the 600 rpm spindle speed, and the feed rate by 0–3 mm/rev) for coir reinforced
polymer composites. Athijayamani et al. [21] predict and evaluate torque and thrust
force in drilling of roselle composite material. The prediction of output parameter is
carried out using regression model (RM) and artificial neural network (ANN) model.
Drill dynamometer was used during the drilling process to calculate thrust strength
and torque. Feed, cutting speed, and drill diameter taken as input parameter while
torque and thrust force taken as output process parameter. Results show the predicted
value of torque and thrust forces are similar to experimental value and ANN model
provides better results than RM model.

From the literature, it has been observed that plenty of work on bamboo and other
fiber-based polymer composites in context with physical, chemical, mechanical, and
few on machinability properties have been done. However, very less work on machinability of short bamboo fiber polymer composites using convetion machining process
has been investegated. Further, machining of SBFRP composites using conventional
machining and non-conventional machining found to be a bit difficult and costly
attempt due to their superior properties such as inhomogeneous and anisotropic [19–
31]. Conventional machining generates excessive heat at the cutting portion and
the heat dissipation rate is decreased due to the relatively low thermal conductivity
of these materials. As a result, it decreases the tool bit life, increases the surface
quality, and decreases the dimensional sensitivity of work materials. To resolve,
many traditional machining processes use cutting fluid during the heating issues in

the machining process. But, the use of various cutting fluids in machining results in various forms of environmental issues or poisonous components in the form of waste which leads to enormous environmental and health issues. To overcome these difficulties, the present chapter used dry machining conditions for machining of SBFRP composites in the CNC drilling process. Also, the present chapter showed the applicability of short bamboo fiber in manufacturing and processing using the traditional machining process under dry machining. However, no work was found in the literature on machining of SBFRP composites and their effect on parameters of machining under dry machining environment using the CNC drilling process. Furthermore, performance of machining strongly depends on its operating parameters and performance parameters. Hence, optimization of machining process is essential. Since the machining process includes many no of input parameters and output parameters and considered to be a multi-criteria decision making problem. It is noted from the literature that optimization of SBFRP composites using MCDM method is hardly available.

2.2 Materials Manufacturing and Machining Details

2.2.1 Manufacturing of SBFRP Composite Using Compression Molding Technique

Raw bamboo as reinforcement materials and environment-friendly material as poly-lactic acid (PLA) pellets as matrix is used for manufacturing/fabrication of composite specimens. First, short fibers/fillers of bamboo are extracted from the raw bamboo

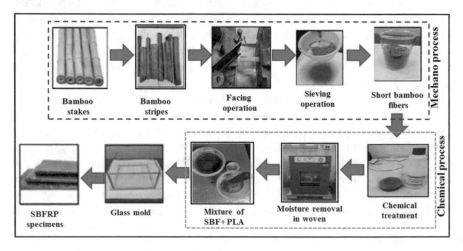

Fig. 2.1 Extraction + manufacturing of SBFP composites using a mechano-chemical process

fibers using the mechano-chemical process (Fig. 2.1). In this process, the outer and inner layers of the bamboo stakes are peeled off first and bamboo culm cuts into small pieces using a hand saw machine. Bamboo culm pieces are then converted to the short bamboo fibers (SBF) via. Facing operation in the lathe machine. Different sized short bamboo fibers are segregated using sieving operation.

Then after that, extracted short bamboo fiber is undergone several chemical treatments (Fig. 2.1) such as alkaline treatment process, i.e., NaOH, acetone treatment, etc. for removing the impurities, lignin, and hemicellulose present in the SBF resulting in greater fiber surface and enhancing the matrix interfacial bonding [32]. In the first chemical treatment, 100 ml of NaOH taken in a beaker and then 5 g of untreated short bamboo fiber are added in the solution. The reaction was carried under a high stirring speed of 3000 rpm for 15 min at room temperature (Fig. 2.1). The processed fibers are then washed by distilled water and acetone solution and then dried in an oven at 70 °C overnight to remove the moisture in surface-treated SBF. Later, treated short bamboo fiber is taken for sample preparation. Primarily, glass sheets cut into rectangle size and attach with glue to form a glass mold. After that, 250 g of PLA pellets are taken in a crucible and placed in a muffle furnace at 220 °C for four hours. Secondary, 12.5 g of treated short bamboo fiber added in the crucible and stirrer for 2 min with help of glass rod. Crucible and glass rod must be properly cleaned with tap water. The tertiary mixture of liquid PLA and SBF poured into glass mold and compresses. Before the pouring of a mixture, silicon spray applied on the inner surface of the mold for easy removal of a composite. The sample of size 150 mm × 100 mm × 10 mm is used for experimentation [32].

2.2.2 Machining Details

The CNC machining is used for machining of short bamboo fiber-reinforced polymer (SBFRP) composites (Fig. 2.2a, b). A series of drill under dry machining conditions was performed with rotating spindle speed (range 500–1200 rpm), feed rate (range from 0 to 5000 mm/min), and thrust force is calculated using a dynamometer. A dynamometer is made using a 40 kg load cell, HX711 module, and unoArduino. The SBFRP composite is attached with the load cell with the help of a C-clamp.

However, the use of various cutting fluids in machining results in various forms of environmental issues or poisonous components in the form of waste which leads to enormous environmental and health issues. Hence, the present chapter used dry machining conditions for machining of SBFRP composites in the CNC drilling process [32, 33]. During the experimentation, cutting speed, feed rate, and point angle are selected as input parameters while thrust force (TF), material removal rate (MRR), and process time (PT) are considered as response parameters. Experiments are performed according to the Taguchi (L18) as tabulated in Table 2.1, and subsequently, corresponding responses are determined (Table 2.2).

During the experiment, three independent parameters (shown in Table 2.1) are changed according to the experimental setup, and a hole of Ø 10 mm is cut [32, 33]).

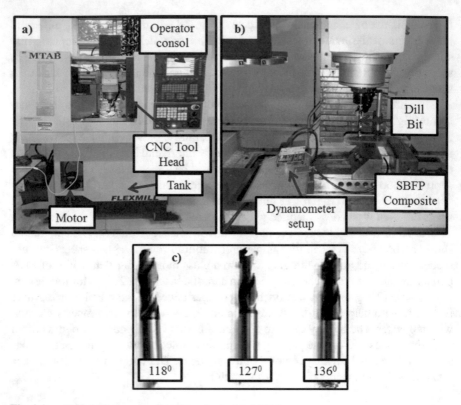

Fig. 2.2 **a** CNC drilling center, **b** tool head setup, **c** drill bits

Table 2.1 Input conditions of CNC drilling center

Input parameters	Symbol	Units	Level 1	Level 2	Level 3
Cutting speed	CS	mm/min	60	120	180
Spindle speed	SS	rpm	500	800	1200
Point angle	PA	deg	118	127	136
Constant parameters					
Shank type	Cylindrical				
Tool material	HSS				
Coolant	Dry machining				

Each experiment is carried out 3 times, and their average of TF, MRR, and PT are taken for the analysis (Table 2.2). The output parameters thrust force, MRR, and process time are evaluated using the following expression:

$$TF = \frac{X \times 9.81}{1000} \tag{2.1}$$

Table 2.2 Experimental results of SBFRP composites in CNC drilling process

Exp No.	% composition	Spindle speed (SS) (rpm)	Cutting speed (CS) (mm/min)	Point angle (PA) (°)	Thrust force (TF) (N)	MRR (mm³/s)	Process time (PT) (s)
1	5%	500	60	118	129.95	10.745	15.01
2		500	120	127	130.31	15.797	10.21
3		500	180	136	136.78	30.319	5.32
4		800	60	127	108.398	11.465	15.41
5		800	120	136	111.891	17.179	9.97
6		800	180	118	115.142	28.148	5.73
7		1200	60	136	69.454	12.482	15.39
8		1200	120	118	72.727	19.658	9.68
9		1200	180	127	80.402	34.079	5.19

$$\mathrm{MRR} = \frac{I_W - F_W}{C_T} \qquad (2.2)$$

where X is a load applied in N, I_W, and F_W are the initial and final weights of SBFRP composite; C_T is cutting time; MRR is material removal rate.

2.3   Results and Discussion

2.3.1   Effects of Process Parameters on Thrust Force

The main effect plots of input parameters like spindle speed, cutting speed, and process time on thrust force are shown in Fig. 2.3a–c. The main effect plot shows that input parameters spindle speed, cutting speed, and point angle are more influenced parameters on thrust force. It is observed from Fig. 2.3a that the process parameter spindle speed increases from 500 to 1200 rpm, and the response parameter thrust force is drastically decreased during the machining of SBFRP composite. The reason behind this is that the amount of heat generation on the hole wall surface raises with higher spindle speed, the generated heat is converted into plastic deformation of polymers, resulting in reduction in thrust force. However, Fig. 2.3b is noticed that the dependent parameter thrust force is significantly increased with increasing the independent parameter cutting speed from 60 to 180 mm/min. This is because the cutting forces between tool and work material are increased with more penetration, resulting in higher thrust force obtained during the machining of SBFRP composite.

Meanwhile, the influence of input parameter point angle on output parameter thrust force as shown in Fig. 2.3c that the response parameter thrust force is slightly

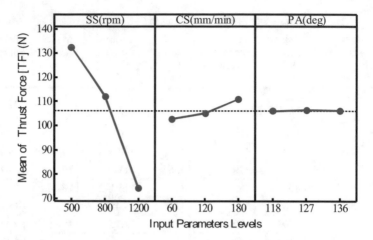

Fig. 2.3 a–c Main effect plot for thrust force versus input parameters levels

increased from 105.8 to 106.4 N with increasing the process parameter point angle
from 118° to 127°. The reason behind this is that the cutting forces are more at
the medium point angle on the workpiece, resulting in higher thrust force. But the
parameter thrust force is decreased from 106.4 to 106 N with an increase in the
process parameter PA from 118° to 136°. This is because; the length of cutting edge
is increased with an increase in point angle. The specimen's bending resistance in
the axial direction was decreased, resulting in the cutting tool easily penetrated into
work material and led to a reduction in thrust force [32]. Based on this analysis, the
optimal combination of process factors SBFRP composites for TF is at spindle speed
(1200 rpm, level 3), cutting speed (60 mm/min, level 1), point angle (118°, level 1).

2.3.2 Effects of Process Parameters on MRR

The impact of input parameters like spindle speed, cutting speed, and point angle on
MRR is shown in Fig. 2.4a–c. It is observed from Fig. 2.4a that the process parameter
spindle speed increased from 500 to 800 rpm, and the value of response parameter
MRR marginally decreased from 19.44 to 18.15 mm³/s. This is because the contact
area between tool and work material decreases with an increase in spindle speed,
resulting in lower MRR. But the response parameter MRR is increased from 18.15
to 19.14 mm³/s with the parameter spindle speed gradually increasing from 800
to 1200 rpm. The reason behind this is that the processing time decreases with an
increase in cutting forces, resulting in higher MRR which is achieved. However, the
independent parameter cutting speed is positively influenced by dependent parameter
MRR as shown in Fig. 2.4b. It shows that the independent parameter cutting speed is
varied from 60 to 180 mm/min, the dependent parameter MRR is gradually increased
from 10.2 to 15.9 mm³/s, and further, it is increased from, 15.9 to 29.2 mm³/s. This is

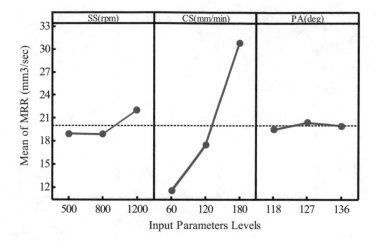

Fig. 2.4 a–c Main effect plot for MRR versus input parameters levels

because the area of contact, cutting forces, and tool penetration are increased between tool and work material, resulting in higher MRR which is achieved.

Moreover, it is clearly observed from Fig. 2.4c that parameter MRR is increasing from 19.5 to 20.4 mm³/s with parameter point angle varying from 118° to 127°. This is because the point angle is at medium, the contact area on the work material is increased, and bending resistance with axial direction is reduced, resulting in higher MRR. But, the response parameter MRR is reduced from 20.4 to 19.8 mm³/s with the process parameter increased from 127° to 136°. The reason behind this is that the cutting forces, contact area, and chip formation are decreased, resulting in lower MRR obtained during the machining of SBFRP composite by using the CNC drilling process [32, 33], and [34]. Based on this analysis, the optimal combination of process factors SBFRP composites for MRR is at spindle speed (1200 rpm, level 3), cutting speed (180 mm/min, level 3), point angle (127°, level 2).

2.3.3 Effects of Process Parameters on Process Time

The effect of input parameters like spindle speed, cutting speed, and point angle on process time of SBFRP composite is illustrated in Fig. 2.5a–c. It is observed from Fig. 2.5a that parameter spindle speed increases from 500 to 800 rpm, and the response parameter process time increases from 10.03 to 10.61 s. The reason behind this is that the cutting forces are decreased with an increase in spindle speed, resulting in higher process time. But response parameter process time drastically decreases from 10.61 to 10.22 s with the process parameter spindle speed increases from 800 to 1200 rpm. This is because discontinuous chip formation is taking place during the higher rate of spindle speed, resulting in reduced machining time or process time.

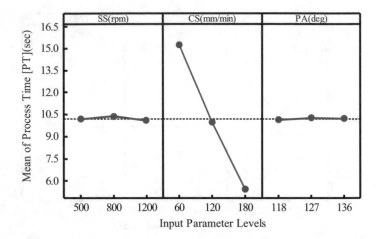

Fig. 2.5 a–c Main effect plot for process time versus input parameters levels

It is clearly noticed that an increase in cutting speed results in a decrease in process time.

With reference to parameter cutting speed, the process time decreases with an increase in cutting speed (Fig. 2.5b) from 60 to 180 mm/rev. Because process time decreases with an increase in cutting speed for all brittle materials and no formation of chip clogging in-between tool and workpiece in the CNC machine [33]. However, from Fig. 2.5c, the dependent parameter process time is gradually increased from 10.14 to 10.27 s; the independent parameter point angle is increased from 118° to 127°. The reason behind this is that the amount of fiber percentage increases with improving the molecular structure of SBFRP composite and increases the resistance between a tool and work material. However, the dependent parameter process time is a marginal decrement from 10.27 to 10.21 s, the independent parameter point angle changing from 127° to 136°. This is because, the contact area on the work material is more and bending resistance with axial direction also more result in, lower the processing time is obtained during the machining of SBFRP composite by using the CNC drilling process [32, 35]. Based on this analysis, the optimal combination of process factors SBFRP composites for process time is at spindle speed (1200 rpm, level 3), cutting speed (180 mm/min, level 3), point angle (118°, level 1).

2.3.4 ANOVA

ANOVA is done to analyze the significant parameters on responses during the machining of SBFRP composites in a dry machining environment using the CNC drilling process. The MINITAB 17.0 software is used for ANOVA and the results of the same are tabulated in Tables 2.3, 2.4 and 2.5. Few testes like the F-ratio test and

Table 2.3 ANOVA for thrust force of SBFRP composites

Source	DF	Adj SS	Adj MS	F-value	P-value
Regression	3	5285.04	1761.68	180.46	0
SS (rpm)	1	852.49	852.49	87.33	0
CS (mm/min)	1	1.49	1.49	0.15	0.712
SS (rpm) × CS (mm/min)	1	4.64	4.64	0.48	0.521
Error	5	48.81	9.76		
Total	8	5333.85			
5% SBFRP	$R^2 = 99.72\%$, R^2 (adj) $= 99.44\%$, R^2 (Pred) $= 98.58\%$				

Table 2.4 ANOVA for MRR of SBFRP composites

Source	DF	Adj SS	Adj MS	F-value	P-value
Regression	3	575.203	191.734	26.16	0.002
SS (rpm)	1	16.30%	16.30%	2.00	0.887
CS (mm/min)	1	42.988	42.988	5.87	0.001
SS (rpm) × CS (mm/min)	1	1.424	1.424	0.19	0.678
Error	5	36.647	7.329		
Total	8	611.85			
5% SBFRP	$R^2 = 94.01\%$, R^2 (adj) $= 90.42\%$, R^2 (pred) $= 68.00\%$				

Table 2.5 ANOVA for process time of SBFRP composites

Source	DF	Adj SS	Adj MS	F-value	P-value
Regression	1	145.731	145.731	1451.22	0
SS (rpm)	1	145.731	145.731	1451.22	0
Error	7	0.703	0.1		
Total	8	146.434			
5% SBFRP	$R = 99.52\%$, R^2 (adj) $= 99.45\%$, R^2 (pred) $= 99.23\%$				

P-test followed by the coefficient of determination (R^2) and adjusted R^2 are calculated to test the enough and fitness of the model [35]. The F-values and P-values indicate the most significant parameter and statistical significance of the input variables by considering the larger F-values and lesser (<0.05) P-values, respectively [34]. The ANOVA for thrust force of SBFRP composites is done using Minitab software considering the experimental data as tabulated in Table 2.2 and corresponding results of ANOVA are tabulated in Tables 2.3. It has been observed from the results (Table 2.3) that F-value for spindle speed (linear parameter) yields the highest with 87.33 and P-value is exactly 0.000. It implies that the dependent parameter spindle speed is the most significant parameter in determining the performance characteristic thrust force for 5% SBFRP composite machining. Although an interaction parameter spindle speed × cutting speed exists, it is most not so significant because of lesser values of F- and P-values. Furthermore, the value of R^2 obtained for thrust force is 99.72% and the adjusted R^2 is 99.44% shows better fitness of the data by presented model [36].

Subsequently, the results of ANOVA for MRR of SBFRP composites are depicted in Table 2.4. The result shows that only linear parameter cutting speed yields higher F-value (i.e., 5.87) and P-values are exactly 0.000 compared to the other parameters. This indicates that only linear parameter cutting speed is the most influencing parameter in determining the performance characteristic MRR during machining. But, none of the other parameters except cutting speed is found significant for SBFRP composite [35]. Moreover, the value of R^2 obtained for MRR is 94.01% and the adjusted R^2 is 90.42% shows better fitness of the data by presented model [34, 35] for both cases.

Similarly, ANOVA for process time is also performed and their results of the same are shown in Table 2.5. The results indicate that only linear parameter spindle speed yields higher F-value (i.e., 1451.22) and P-values is exactly 0.000 for % SBFRP composite machining in the CNC drilling process [35]. Hence, the most effective parameters in all the cases were found to be parameter cutting speed and spindle speed for SBFRP composites. Additionally, the value of R^2 obtained for process time is 99.52% and the adjusted R^2 is 99.45% shows better fitness of the data by presented model [34] for both cases.

2.3.5 Empirical Model

In this section, the advancement/development of empirical models for every one of the yield boundary or output parameters, for example, thrust force, MRR, and process time are examined. The development of empirical models is done by employing multiple regression analysis in Minitab programming software and comparing plots produced additionally been created. Each empirical model of thrust force, MRR, and process time comprises of a lot of conditions Eqs. (2.3–2.5)] of linear and association terms of input parameters (e.g., spindle speed, cutting speed, and point angle) [36]

$$\text{TF}(N) = 172.78 - 0.08980 \times \text{SS} + 0.0225 \times \text{CS} + 0.00051 \times \text{CS} \times \text{SS} \quad (2.3)$$

$$\text{MRR}(\text{mm}^3/\text{s})) = -0.33 + 0.00124 \times \text{SS} + 0.1371 \times \text{CS}$$
$$+ 0.000028 \times (\text{SS} \times \text{CS}) \quad (2.4)$$

$$\text{PT}(\text{s})) = 20.069 - 0.08214 \times \text{SS} \quad (2.5)$$

Further, empirical models (Eqs. 2.3–2.5) for thrust force, MRR, and process time are employed for optimum prediction of the output parameters for the CNC drilling process considering different inputs (spindle speed, cutting speed, and point angle) conditions on machining of SBFRP composites. The predicted output parameters are further utilized or tested in its normality via normal probability plots. The normal probability plots (Fig. 2.6) shows that all the points of the response thrust force, MRR, and process time are closer to the straight lines and hence follow the normality of the data well within the confidence interval of 95% [36].

2.3.6 Modeling and Optimization

Multi-objective optimization of CNC drilling parameters on machining of SBFRP composites under dry machining environment is done using an integrated multi-criteria decision making (MCDM) based method, i.e., entropy- multi-criteria ranking analysis (MCRA). Here, entropy [37, 38] is used for the extraction of particular arrangement weights of the CNC responses while MCRA [39, 40] is used for ranking and optimization. In optimization, spindle speed, cutting speed, and point angle parameters as input and thrust force, MRR, and process time as output parameters are considered. First, the design of the decision matrix is carried out based on the Taguchi L_9-OAusing Eq. (2.6) and the results of the shown in Table 2.2.

$$D_{ij} = \begin{bmatrix} & C_1 & C_2 & \dots & C_n \\ A_1 & Y_{11} & Y_{12} & \dots & Y_{1n} \\ A_2 & Y_{21} & Y_{22} & \dots & Y_{2n} \\ \dots & \dots & \dots & \dots & \dots \\ A_m & Y_{m1} & Y_{m2} & \dots & Y_{mn} \end{bmatrix} \tag{2.6}$$

where D_{ij} is the decision matrix contains response values of ith alternatives on jth criterion, i.e., Y_{11}, Y_{12}, Y_{mn}; C_1, C_2, ..., C_n represents the number of criteria or no. of response parameters; A_1, A_2, ..., A_m is a no. of alternatives or experiments.

In the second step, normalization of the decision matrix is done using Eqs. (2.7 and 2.8) based on the type of output parameters. This step converts the different measurement data of output parameters into comparable data in the range of 0–1. Here, output parameter MRR is considered as beneficial criteria while thrust force and process time as non-beneficial criteria. The result of the normalized matrix is tabulated in Table 2.6.

Beneficial criterion

$$[B_{ij}]_{m \times n} = \left[\frac{Y_{ij}}{\max \sum_{i=1}^{m} Y_{ij}^2} \right]_{m \times n} \tag{2.7}$$

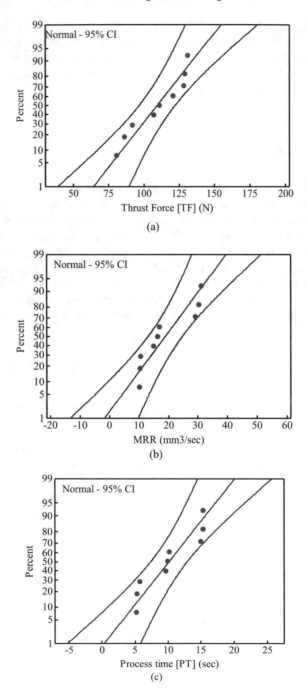

Fig. 2.6 Probability plot for: **a** Thrust force (TF), **b** MRR, **c** process time (PT) of SBFRP composites

Table 2.6 Normalized matrix

Exp No.	TF in (N)	PT in (s)	MRR in (mm³/s)
1	0.1361	0.1633	0.0841
2	0.1364	0.1111	0.0955
3	0.1432	0.0579	0.0766
4	0.1135	0.1677	0.0693
5	0.1172	0.1085	0.1039
6	0.1206	0.0623	0.1702
7	0.0727	0.1674	0.0755
8	0.0761	0.1053	0.1189
9	0.0842	0.0565	0.2061

Non-beneficial criterion

$$\left[NB_{ij}\right]_{m \times n} = \left[\frac{Y_{ij}}{\min \sum_{i=1}^{m} Y_{ij}^2} \right]_{m \times n} \tag{2.8}$$

where B_{ij} and NB_{ij} denotes the normalized evaluation matrix for beneficial and non-beneficial criterion and max (Y_{ij}) indicates the maximum response value of the criterion for beneficial and min (X_{ij}) is the minimum response value criterion for non-beneficial.

Third, particular arrangement weights of response parameters are calculated based on the entropy method [37, 38] using Eqs. (2.10 and 2.11) followed by a degree of divergence for each of the output parameters, i.e., thrust force, MRR, and process time. The degree of divergence denotes the distance of each output parameter w.r.t comparable data and tells the closeness of the data. Based on the degree of divergence values, precise priority weights of the responses are calculated using the following expressions and corresponding weights are depicted in Table 2.7.

$$\delta_j = -c \sum_{i=1}^{m} B_{ij}, NB_{ij}, \ln(B_{ij}), \quad \ln(NBi_j) \ j = 1, 2, \ldots, n. \tag{2.9}$$

$$d_j = \left| 1 - \delta_j \right|. \tag{2.10}$$

$$\zeta_j = \frac{d_j}{\sum_{j=1}^{n} d_j}. \tag{2.11}$$

Table 2.7 Priority weights of CNC drilling output parameters

% composition	TF in (N)	PT in (s)	MRR in (mm³/s)
5%	0.3283	0.3373	0.3344

where c is a constant term and calculated as $(k = 1/\ln m, 0 \le d_j \le 1)$, ζ_j are the particular arrangement weights of jth criteria.

The arrangement of weights (Table 2.7) observed that the output parameter MRR (0.3283) and process time (0.3373) yields higher weights for SBFRP composites. This signifies that parameter MRR and process time are very much significant in the input parameters of the CNC drilling process. After the priority weight calculation, the formulation of the weighted decision matrix is done using Eq. (2.12). In this step, weights of the output parameters, i.e., thrust force, MRR, and process time are multiplying with each of the response values of the decision matrix.

$$\omega_{ij} = \left[B_{ij}\right]_{m \times n} \text{or} \left[NB_{ij}\right]_{m \times n} \times \zeta_j \tag{2.12}$$

where ω_{ij} denotes the weighted decision matrix; ζ_j represents the weights of each criterion (thrust force, MRR and process time), and B_{ij} and NB_{ij} are the normalized decision matrices.

Next, assessment of closeness index (CI) values for each and every alternative, i.e., exp. no is carried out using Eqs. (2.13–2.15). These equations convert the multi-objective or multi-response to a single objective optimization problem taking into account beneficial and non-beneficial parameters [39, 40], and [41].

$$\alpha = \sum_{i=1}^{b} \omega_{ij} \tag{2.13}$$

$$\beta = \sum_{i=b+1}^{n} \omega_{ij} \tag{2.14}$$

$$CI = \alpha - \beta \tag{2.15}$$

where α and β denotes the summation of weighted response values (beneficial and non-beneficial) of jth criteria (thrust force, MRR and process time), b is the no. of beneficiary criteria, n indicates the total no. of criteria, and CI is the closeness index values.

Last, the ranking of alternatives, i.e., exp. no is finished based on CI values. The choice with greater CI gives optimal operating conditions to get higher performance for the system compared to the other experiment number [41]. The outcomes of the ranking are tabulated in Table 2.8. The results show that exp. No. 9 yields the highest ranking in both cases compared to the other exp. runs. This shows that input parameters, i.e., spindle speed (1200 rpm), cutting speed (180 mm/min) and point angle (127°) gives the optimal output parameters, i.e., thrust force (80.402 N), MRR (34.079 mm³/s), process time (5.19 s), respectively.

The result shows that, for 5% SBFRP composites in the CNC drilling process, Exp. No. 9 achieved the highest attainment values among the other runs. The closeness (CI) value for optimal setting obtained is 0.0693. This shows the closeness (CI) value among the other experiments. The following optimal setting yields the best

Table 2.8 CI values and ranking of the alternatives of CNC drilling process

Exp No.	CI values	Rank
1	0.0102	8
2	0.0645	3
3	0.0399	6
4	0.0714	2
5	0.0473	4
6	0.0030	7
7	0.0470	5
8	0.0005	9
9	**0.0693**	1

combination of process variables are spindle speed (1200 rpm, level 3), cutting speed (180 mm/min, level 3), and point angle (127°, level 3) gives the optimal output parameters, i.e., thrust force (80.402 N), MRR (34.079 mm^3/s), process time (5.19 s). The optimal setting obtained using entropy-MCRA method provides the most optimal values for CNC drilling processes which have less influence on the performance of CNC drilling process during the machining of SBFRP composites. Also optimal setting creates less environmental impacts (because lower process time, lower thrust process and higher MRR as well as produce no environmental pollution due to the use of dry machining conditions). Also it improves the quality of the product, reduces the final product costs, and improves the machining efficiency of the CNC drilling process [41].

Therefore, the optimal values such as MRR, thrust force, and process time under dry machining result in the CNC drilling process toward the green manufacturing process. Hence, it is recommended that optimal setting, i.e., spindle speed (1200 rpm), cutting speed (180 mm/min), and point angle (127°), is used in the CNC drilling for effective machining of SBFRP composites under tap water. Furthermore, the work suggested that the entropy-MCRA method is used as a systematic framework model for optimal selection of process parameter for effective machining of SBFRP composites under dry machining in the CNC drilling process in green manufacturing environments.

2.3.7 Confirmatory Analysis

Additionally, the justification of results obtained from the entropy-MCRA method is done through the confirmatory analysis. For this, an optimal setting [spindle speed (1200 rpm, level 3), cutting speed (180 mm/min, level 3) and point angle (127°, level 3)] obtained using the proposed method is considered and the same being re-tested in CNC drilling process. It has been observed from the results (Table 2.9)

Table 2.9 Confirmatory tests for CNC drilling process of SBFRP composites

Input conditions	Output parameters	Experimental results	Confirmatory tests results
Exp. No. 9	*5% SBFRP composite*		
SS (1200 rpm), CS (180 mm/rev) and PA (127°)	TF (N)	80.402	81.761
	MRR (mm³/s)	34.079	33.876
	PT (s)	5.19	5.34

that the confirmatory test results found to be comparable and acceptable with the experimental results for the selected optimal setting [41, 42].

2.4 Summary

The present chapter aimed at manufacturing and processing of SBFRP composites in the CNC drilling process. For this fabrication of SBFRP composites using compression molding technique and evaluation of processing parameters (spindle speed, cutting speed and point angle of drill bit) on thrust force, MRR and process time in processing. ANOVA and multi-regression analysis are employed to study the statistical significance. The optimal setting for CNC drilling is found through fuzzy-MOOSRA method. Based on the experimental design, results, ANOVA, parametric and optimization results the following conclusions are pinched.

- CNC milling process is found feasible in machining of short bamboo-based polymer composites under its operating parameters.
- Compression molding technique is more effective in production of fiber-based polymer composites compared to the hand layup process.
- Based on the parametric analyses, the optimal combination of process factors short bamboo fiber-reinforced polymer composites for thrust is obtained at spindle speed (1200 rpm, level 3), cutting speed (60 mm/min, level 1), point angle (118°, level 1). MRR is obtained at spindle speed (1200 rpm, level 3), cutting speed (180 mm/min, level 3), and point angle (127°, level 2). Process time is obtained at spindle speed (1200 rpm, level 3), cutting speed (180 mm/min, level 3), and point angle (118°, level 1).
- Based on the ANOVA, the parameter, spindle speed yields greater effect on thrust force, MRR and process time in processing of SBFRP composites. Further, ANOVA and empirical models for process time, thrust force, and MRR of short bamboo fiber-reinforced polymer composites are statistically significant and fit the data well enough with the experiment results with the 95% confidence level.
- Multi-response parameter optimization is done using Fuzzy-MCRA method. The optimal setting obtained via proposed method is at spindle speed (1200 rpm, level 3), cutting speed (180 mm/min, level 3) and point angle (127°, level 3) which

provides optimal responses as thrust force (80.402 N), MRR (34.079 mm^3/s), process time (5.19 s), respectively.

- Additionally, the justification of results obtained from the entropy-MCRA method is done through the confirmatory analysis. It has been observed from the results that the confirmatory test results found to be comparable and acceptable with the experimental results for the selected optimal setting.

In view of the above observations, the short bamboo fiber-reinforced polymer composites give better execution in the CNC drilling process. The parametric settings got from the entropy-MCRA technique can be utilized as the ideal setting for the CNC drilling during the machining of short bamboo fiber-reinforced polymer composites. Moreover, the created empirical model for CNC drilling responses can be implemented in an organized structural model for prediction of process time, thrust force, and MRR for all the cases.

References

1. D. Chandramohan, Studies on natural fiber particle reinforced composite material for conservation of natural resources. Adv. Appl. Sci. Res. **5**(2), 305–315 (2014)
2. K. Georgios, S. Arlindo, F. Mihail, Green composites: a review of adequate materials for automotive applications. Compos. Part B-Eng. **44**, 120–127 (2013)
3. K.C.M. Nair, R.P. Kumar, S. Thomas, S.C. Schit, K. Ramamurthy, Rheological behavior of short sisal fiber-reinforced polystyrene composites. Compos. Part A Appl. Sci. Manuf. **31**, 1231–1240 (2000)
4. R.B. Raj, B.V. Kokta, F. Dembele, B. Sanschagrain, Compounding of cellulose fibers with polypropylene: effect of fiber treatment on dispersion in the polymer matrix. J. Appl. Polym. Sci. **38**, 1987–1996 (1989)
5. N.C. Billingham, *Plastics Additives*, ed. by R. Gachter, H. Muller, 3rd edn (Hanser verlag, munich, 1990). Polym. Int. **25**(4), 260–260
6. F.P. La-Mantia, M. Morreale, Green composites: a brief review. Compos. Part A Appl. Sci. Manuf. **42**, 579–588 (2011)
7. H.P.S. Abdul Khalil, A.H. Bhatt, A.F.I. Yusra, Green composites from sustainable cellulose nanofibrils: a review. Carbohydr. Polym. **87**, 963–979 (2012)
8. J. Markarian, Additive developments aid growth in wood-plastic composites. Plast. Addi. Compou. **4**, 18–21 (2002)
9. G. Pritchard, Two technologies merge: wood-plastic composites. Plast. Addi. Compou. **6**, 18–21 (2004)
10. Y. Yu, R. Liu, Y. Huang, F. Meng, W. Yu, Preparation, physical, mechanical, and interfacial morphological properties of engineered bamboo scrimber. Constr. Build. Mater. **157**, 1032–1039 (2017)
11. S. Huda, N. Reddy, Y. Yang, Ultra-light-weight composites from bamboo strips and polypropylene web with exceptional flexural properties. Compos. Part B Eng. **43**(3), 1658–1664 (2012)
12. A.C. Manalo, E. Wani, N.A. Zukarnain, W. Karunasena, K.T. Lau, Effects of alkali treatment and elevated temperature on the mechanical properties of bamboo fibre–polyester composites. Compos. Part B: Eng. **80**, 73–83 (2015)
13. J.K. Huang, W.B. Young, The mechanical, hygral, and interfacial strength of continuous bamboo fiber reinforced epoxy composites. Compos. Part B: Eng. **166**, 272–283 (2019)

14. G.D. Babu, K.S. Babu, B.U.M. Gowd, Effects of drilling parameters on delamination of hemp fiber reinforced composites. Int. J. Mech. Eng. Res. Dev. **2**, 1–8 (2012)
15. P.N.E. Naveen, M. Yasaswi, R.V. Prasad, Experimental investigation of drilling parameters on composite materials. J. Mech. Civ. Eng. **2**, 30–37 (2012)
16. T.B. Yallew, P. Kumar, I. Singh, A study about hole making in woven jute fabric-reinforced polymer composites. Proc. Inst. Mech. Eng. Part L: J. Mater.: Des. App. **230**(4), 888–898 (2016)
17. D. Wang, P.Y. Onawumi, S.O. Ismail, H.N. Dhakal, I. Popov, V.V. Silberschmidt, A. Roy, Machinability of natural-fibre-reinforced polymer composites: conventional versus ultrasonically-assisted machining. Compos. Part A: App. Sci. Manuf. **119**, 188–195 (2019)
18. D. Chandramohan, K. Marimuthu, Drilling of natural fiber particle reinforced polymer composite material. Int. J. Adv. Eng. Res. Stud. **1**(1), 134–145 (2011)
19. G. Dilli Babu, K. Sivaji Babu, U.M. Gowd, Effect of machining parameters on milled natural fiber-reinforced plastic composites. J. Adv. Mech. Eng. **1**, 1–12 (2013)
20. S. Jayabal, U. Natarajan, Drilling analysis of coir-fibre-reinforced polyester composites. Bull. Mater. Sci. **34**(7), 1563–1567 (2011)
21. M. Zampaloni, F. Pourboghrat, S.A. Yankovich, Kenaf natural fiber—a discussion on manufacturing problems and solutions. Compos. Part A Appl. Sci. Manuf. **38**(6), 1569–1580 (2007)
22. A. Mizobuchi, H. Takagi, T. Sato, J. Hino, Drilling machinability of resin-less green composites reinforced by bamboo fiber. WIT Transac. Built. Envir. **97**, 186–194 (2008)
23. P.K. Bajpai, I. Singh, Drilling behavior of sisal fiber-reinforced polypropylene composite laminate. J. Reinf. Plast. Compo. **32**(20), 1569–1576 (2013)
24. V. Sridharan, N. Muthukrishnan, Optimization of machinability of polyester/modified jute fabric composite using grey relational analysis (GRA). Proc. Eng. **64**, 1003–1012 (2013)
25. K.K. Chawla, Ch.1 Introduction, in K.K. Chawla, *Composite Materials Science and Engineering* (Springer, New York, 2013). (1), 105–106
26. R. Teti, Machining of composite materials. CIRP J. Manuf. Sci. Technol. **51**(2), 611–634 (2002)
27. S. Abrate, Machining of composite materials, in *Composites Engineering Handbook*, ed. by P.K. Mallick (Marcel Dekker, New York, 1997), pp. 777–809
28. M.K. Surppa, Aluminum matrix composites: challenges and opportunities. Sadhana **28**, 319–334 (2003)
29. A.W. Momber, R. Kovacevic, *Modelling of Abrasive Water—Jet Cutting Processes, Principle of Abrasive Waterjet Machining* (Spinger, London, 1998), pp. 163–194
30. W. Konig, S. Rummenholler, Technological and industrial safety aspects in milling FRP. ASME Machi. Adv. Compo. **45**(66), 1–14 (1993)
31. S.P. Sivapirakasam, J. Mathew, M. Surianarayanan, Multi-attribute decision making for green electrical discharge machining. Exp. Syst. Appl. **38**, 8370–8374 (2011)
32. R.M. Jagadish, A. Ray, Investigation on mechanical properties of pineapple leaf–based short fiber–reinforced polymer composite from selected Indian (northeastern part) cultivars. J. Thermoplast. Compos. Mater. **33**(3), 324–342 (2018)
33. B.S. Jagadish, A. Ray, Prediction and optimization of process parameters of green composites in AWJM process using response surface methodology. Int. J. Adv. Manuf. Technol. **87**, 1359–1370 (2016)
34. J. Kechagias, G. Petropoulos, Application of Taguchi design for quality characterization of abrasive water jet machining of TRIP sheet steels. Int. J. Adv. Manuf. Technol. **62**, 635–643 (2012)
35. S. Jayabal, U. Natarajan, Drilling analysis of coir-fibre-reinforced polyester composites. Bull. Matter. Sci **34**(7), 1563–1567 (2011)
36. R. Kumar, K. Kumar, S. Bhowmik, Mechanical characterization and quantification of tensile, fracture and viscoelastic characteristics of wood filler reinforced epoxy composite. Wood Sci. Technol. **52**, 677–699 (2018)
37. W. Konig, C. Wulf, H. Willerscheid, Machining of fibre reinforced plastics. CRIP-Aann. Manuf. Technol. **34**, 537–548 (1985)

38. M.K. Pradhan, Estimating the effect of process parameters on MRR, TWR and radial overcut of EDMed AISI D2 tool steel by RSM and GRA coupled with PCA. Int. J. Adv. Manuf. Technol. **68**, 591–605 (2013)
39. M. Emma, S. Kieran, M. Vida, An assessment of sustainable housing affordability using a multiple criteria decision making method. Omega **41**(2), 270–279 (2013)
40. H. Caliskan, B. Kursuncu, C. Kurbanoglu, S.Y. Guven, Material selection for the tool holder working under hard milling conditions using different multi criteria decision making methods. Mater. Des. **45**, 473–479 (2013)
41. W.K.M. Brauers, E.K. Zavadskas, The MOORA method and its application to privatization in a transition economy. Control Cybern. **35**(2), 445–469 (2006)
42. B.S. Jagadish, K. Gupta, Modeling and optimization of advanced manufacturing Processes, in *Manufacturing and Surface Engineering*, 1st edn (Springer International Publishing, 2019), p-1–74, https://doi.org/10.1007/978-3-030-00036-3

Chapter 3
Manufacturing and Processing of Hemp Filler-Based Polymer Composite

Abstract Today, due to the demand of newer materials and increase of sustainability issues, polymer industries are started to use natural fiber-based polymer composites or green composites. Before actual use, manufacturing and processing of these composites is essential. Processing of these composites is found bit difficult and costly attempt and produce serious environmental and health issues. To overcome these problems, in the present chapter, an attempt is made on manufacturing and processing of hemp fiber-based reinforced polymer (HFRP) composites in green processing environment. Convectional hand layup method is used for manufacturing while conventional CNC drilling process with tap water as dielectric fluid is used for processing of HFRP composites. Main emphasis is given toward the processing of HFRP composites under green machining environment considering (such as thrust force, MRR and process energy) as a processing parameters followed by parametric analysis, ANOVA, multi-regression analysis and optimization by MOORA method. Moreover, processing data of each of the analysis are summarized and optimal results for HFRP composites are suggested.

Keywords HFRP composite · Green machining · Parametric analysis · ANOVA · Regression analysis · Optimization

3.1 Introduction

Hemp fiber is commonly called as plant-based fiber and most ancient farming of the plant in the USA from the seventeenth to the mid-twentieth centuries. Karus and Vogt [1] studied cultivation, method of extraction, and product lines of hemp plants, and these fibers are widely used in different sectors like paper and textile industries in the EU. Hemp (commonly named as *cannabis sativa*) is a modern farming crop in India at an approximated area of 10,000 hectares in the 2018 census. Mostly, In India Uttarakhand, Uttar Pradesh, Himachal Pradesh, and Karnataka are the two states where that government allowed for the cultivation of hemp. As per IIHA (International industrial hemp association) regulates the hemp grown in India can be processed only for fibers. Hemp (*Cannabis sativa*) and marijuana (*Cannabis*

© The Author(s), under exclusive license to Springer Nature Switzerland AG 2021   39
Jagadish and S. Bhowmik, *Manufacturing and Processing of Natural Filler Based Polymer Composites*, SpringerBriefs in Applied Sciences and Technology,
https://doi.org/10.1007/978-3-030-65362-0_3

Fig. 3.1 Hemp fiber: **a** plant, **b** stems

indicia) are two popular variety species of the cannabis plant. The major advantages in the hemp plant (Fig. 3.1a)are every part of the plant is useful, viz. fiber (bast), stake, seed, flower, and lends, out of which bast is a most beneficial part of the hemp plant. The fibril extracted from the stem of the hemp plant can be utilized in many applications like clothing, construction, paper, health, foods, body care, textiles, and bio-fuels so that it is eco-friendly fiber. Now a day's hemp is widely used in the automobile industries because of its high impact, tensile, low moisture absorption, and flexural properties. Hemp generally creamy, gray, green, and fiber is grown (Fig. 3.1b) approximately between 0.91 m (3 ft) and 4.6 m (15 ft) long, based on the method of extraction of the fibrils from the stem. The chemical composition of hemp fibers consists of cellulose (60–70%), hemicelluloses (15–20%), lignin (2–4%), and pectin (1–2%) and fat and wax (1–2%). Because of the high content of cellulose, degree of crystallinity, hemp fibers are selected as reinforcement material in the present work. Whereas synthetic fibers such as glass fibers are also been used as reinforcement material, but these fibers offer a high density, costly, non-biodegradable, and incur various health problems.

Instead, hemp fibers are biodegradable, less in density, and economically cheap and have larger industrial applications compared to the other fibers (jute, coir, flax, coconut, sisal, and banana, bamboo, bagasses, rice husk, etc.). Meanwhile, some extent of work has been done on the hemp fiber-based composites with their physical and mechanical properties. Mwaikambo and Ansell [2] present the chemical treatment on natural fibers (hemp, sisal, jute, and kapok) with NaOH solution. The result shows that it increases the crystallinity index, thermal stability, and improves the mechanical properties. Kostic et al. [3] explored the chemical treatment of hemp fibers with NaOH (5 and 18% w/v). The results observed that chemically modified hemp fibers having good mechanical properties, less moisture absorption with increased flexibility. Suhara and Mohini [4] discussed the mechanical and water ingestion properties of hemp fiber-reinforced polymer (HFRP) composites and half breed hemp-glass fiber composites. The outcome of the work indicated that, if the strands long haul blurring in water diminishes the mechanical properties of hemp fiber reinforced polymer composites. Thygesen et al. [5] discussed the chemical characterization of hemp sheaves. Fibers are chemically treated by NaOH, Ca(OH)$_2$ and

EDTA and its characterization determined by using FTIR spectroscopy and thermal gravimetric and DSC method. Results reveals that hemicellulose, lignin and waxes are partially removed, and also changes occur in the surface of hemp sheaves.

In addition, the past examination, for example, Gupta and Rao [6] investigated mechanical properties and water absorption of various wt% of hemp fiber reinforced polymer composites. The results showed that the combination of hemp fibers good among other natural fibers, and gives high strength to weight ratio. Somashekar [7] investigated on HFRP composites and explored the mechanical properties like impact and flexural strengths. The result shows that flexural and impact strengths were increasing with further addition of fiber loading and obtained higher flexural and impact strength of HFRP composites compared to the other natural fibers. Naveen et al. [8] investigated the mechanical properties and water absorption of various weight % of sisal/hemp fiber-reinforced polymer composites. The results showed that the combination of sisal/hemp fibers good among other natural fibers, and gives high strength to weight ratio. Gilles Sebert et al. [9] have done the mechanical properties (impact and flexural strengths) analysis of HFRP composites. Work reported that fiber loading plays a significant role in enhancement of flexural and impact strengths of HFRP composites.

Apart from the physical, chemical, and mechanical characterization, HFRP composites undergo various machining operations known as secondary operations in order to get required engineering products. The most commonly used machining operations are drilling, milling, trimming, etc. Machining of HFRP composites can be done by conventional and non-conventional machining [10–14]. In this regard, few attempts have been done on machining of other natural fiber-based polymer composites like Jayabal et al. [15] investigated the machining behavior of coir fiber-reinforced polymer composites on drilling parameters using non-linear mathematical equations (Taguchi approach). Results show that the obtained optimum values will give the lesser power consumption, cutting forces and reduced worn out of the tool, and also achieving good and high precision holes. Patel et al. [16] examined the effect of speed, feed and drill point angle on thrust force and delamination factor during the machining of banana fiber-based polymer composites. Using mathematical equations (i.e., ANOVA approach) the optimums, drill point angle, speed and feed is obtained for the composites. The results showed that the optimal value for thrust force and delamination factor at entry and exit is at $90°$ as point angle, 1000 rpm as cutting speed and 0.1 mm/rev as feed is established. Maleki et al. [17] performed machinability of jute fiber-based polymer composites in drilling process. Drill bit types along with drilling parameter were used for study and found that drill bit types have a greater effect on the performance of drilling process in jute fiber-based composites. Kumar et al. [18] studied the drilling parameters and their optimum values for achieving lower thrust forces and minimum torque with better quality of the drilled holes on banana fiber-based polymer composites. Ramesh et al. [19] emphasize on machining characteristics of the hybrid composite by varying the cutting speed, feed rate, tool diameter. The feed rate of drill bit causes damage like matrix burning, cracking, fiber pull-out, and delamination around the hole. Maximum thrust force, torque, and high feed rate cause delamination and also more number of dislocations of fibers at higher feed rates. As a result, the delamination is maximum at high feed rates. Manickam and Gopinath [20] studied on drilling behavior of sisal-glass hybrid fiber composites. Drilling parameters, namely feed rate, spindle speed and drill diameter as input variables and thrust force and quality of hole as output variables for the study. Work suggested that greater spindle speed, lower feed rate and small drill diameter are preferred for drilling of hybrid composites. Chandramohan and Marimuthu [21] studied the drilling behavior of sisal fiber-based polymer composite considering the drill bit diameter and feed rate on torque and thrust force. Work reported that torque and thrust forces increased with the drill bit diameter and feed rate during machining. Prabhu et al. [22] evaluated machining behavior of banana fiber-based polymer composites in AWJM process. Work considered traverse speed and standoff distance on MRR and surface roughness. Work reported that standoff distance is more influenced parameter for MRR and surface roughness. Similar study done by Selvam et al. [23] using AWJM of hybrid composite. They also reported that standoff distance along with water pressure plays a major role in achieving the good quality cut in AWJM. Work also studied the effect of input parameters on kerf taper and suggested the optimal setting for getting the quality output and good surface finish.

From the literature it is observed that using conventional and non-conventional machining processes, machining of HFRP/other natural fiber-based composites found a bit difficult and costly attempt due to their superior properties such as inhomogeneous and anisotropic. Also, the conventional machining process generates excessive heat at the cutting zone due to the less thermal conductivity of natural fibers, and the heat dissipation rate is decreased. As a result, it decreases the tool bit life, improves the surface quality, and decreases the dimensional sensitivity of work materials [15–20, 24]. To resolve, many conventional machining processes use cutting fluid during the heating issues in the machining process. Cutting fluids in the machining process produces an ample amount of hazardous components in the form of fluids (liquids and gases) and solid forms that make the environmental pollution in the machining process. These poisonous components can enter the body of the operator through injection, inhalation, and skin contact. To overcome these issues, the present chapter attempted to use CNC drilling process with tap water as dielectric fluid during machining of HFRP composites. This is because tap water in machining does not produce any dangerous gases in comparison with other dielectric fluids and is more secure and eco-friendly during the machining process and makes the CNC drilling process toward the greener manufacturing process. However, processing/machining of HFRP composites using the CNC drilling process under the green machining environment has not been done so far in the literature.

3.2 Materials Manufacturing and Machining Details

3.2.1 Manufacturing of HFRP Composites Using Hand Layup Technique

In this section, manufacturing steps hemp fiber-based polymer composites is discussed. The detail steps of extraction and fabrication steps of hemp fiber-based reinforced polymer composites is depicted in Fig. 3.2a, b. Figure 3.2a shows the extraction of hemp fiber from the raw hemp stems while Fig. 3.2b shows the complete fabrication steps of hemp fiber-reinforced polymer composites. Before preparation of samples, extraction of hemp fiber is done. First, raw hemp stems (cannabis sativa stem) are collected (Fig. 3.1a, b). Second, hemp stems are washed and drowned in water for duration of one week to allow for bacterial fermentation to take place. After fermentation, wet stems are removed from the water and dried into the sunlight for a period of 7 days within the cropland, then after peeled fibers from the stem and washed with clean distilled water. Third, retting process is carried out to remove the top surface elements of the hemp stems and hemp fibers which are obtained. Later these fibers are then dried for 2–3 days in sunlight for removal of moisture and water content.

On the other hand, complete fabrication steps of hemp fiber-reinforced polymer composites are explained in Fig. 3.2b. Here, extracted long fibers are converted into

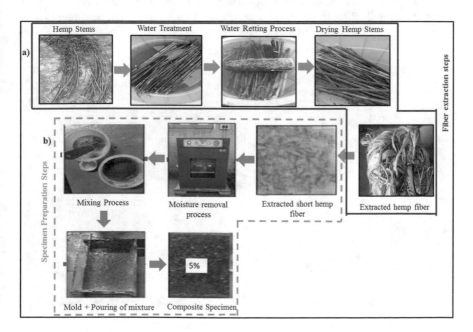

Fig. 3.2 a, b. Extraction + fabrication steps of HFRP composites

short fibers through cutting process fibers first. Second, various chemical treatments (NaOH and acetone treatment) are done with short hemp fiber to remove the impurities and some lignin elements form the surface of hemp fibers. After the treatment, moisture presences in the treated fibers are kept in vacuum oven around 12 h. to remove the moisture. Later, hemp fiber cut into small pieces and converted into powder form with particle size is 300 μm and density 0.623 g/cc and reinforced material weigh percentage is 5% in matrix material, i.e., epoxy resin (Araldite LY 556) with the density of 1.26 g/cc and corresponding hardener (HY 951) in the ratio of 100:10 (by weight). Then after epoxy filled with HFRPs is stirred mechanically and poured gradually into the vacuum chamber made of glass, dimension 110 mm × 45 mm × 8 mm as shown in Fig. 3.2b. To extract the moisture present in the mixture is dried for two days at 28–30 °C. Finally, specimens (Fig. 3.2b) of size 110 mm × 45 mm × 8 mm are taken for machining [24].

3.2.2 Machining Procedure

The drilling process on HFRP composites is conducted on a three-axis vertical CNC drilling machine (Fig. 3.3a, b). During the experiments, a work specimen is fixed with proper adjustable fixtures on the worktable of the CNC drilling machine as shown in Fig. 3.3b.

The drilling operation is carried out using a drill bit with a diameter (ϕ8 mm) to minimize the drilling force on the specimens. Two independent parameters such as spindle speed (SS) and feed rate (FR) with tap water as dielectric fluid as a

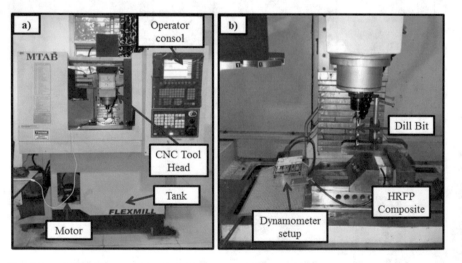

Fig. 3.3 **a** CNC drilling machine, **b** drill bit setup of CNC machine

constant parameter are considered as input parameters while thrust force (TF), material removal rate (MRR), and process energy (PE) are considered as output parameters. HSS tool with diameter ($\phi 8$ mm) with cylindrical shank is used as a tool and tap water is employed as a dielectric fluid. The main reason for using tap water is for the ecological issues because tap water is good working environment and does not produce any kind of harmful gases during the machining in comparison with other fluids. This results in less environmental and health issues and makes the CNC drilling process toward the green manufacturing. The input parameters [17] with their levels and experimental results are depicted in Tables 3.1 and 3.2, respectively.

In experiments, three independent parameters (Table 3.1) are varied according to the Taguchi (L_9-OA) design, and a total of nine circular holes of $\phi 8$ mm is drilled using a CNC drilling machine. Each experiment is carried out 3 times and their average of thrust force, MRR, and process energy is taken for the analysis (shown in Table 3.2). The output parameters thrust force, MRR, and process energy are calculated using the following empirical relations.

$$TF = \frac{X \times 9.81}{1000} \tag{3.1}$$

Table 3.1 CNC drilling process parameters

Input parameters	Symbol	Units	Level 1	Level 2	Level 3
Feed rate	FR	mm/rev	60	120	180
Spindle speed	SS	rpm	500	800	1200
Constant parameters					
Shank type	Cylindrical				
Tool material	HSS				
Coolant	Tap water				

Table 3.2 Machining results of HFRP composites in CNC drilling process

Exp. No.	% Composition	Input parameters		Dielectric fluid	Output parameters		
		SS (rpm)	FR (mm/rev)		TF (N)	MRR (mm³/s)	PE (J)
1	5%	500	60	Tap water	153.39	30.900	7.487
2		500	120		110.54	35.325	4.715
3		500	180		84.69	55.063	2.605
4		800	60		149.32	38.429	7.538
5		800	120		102.67	53.369	4.508
6		800	180		73.02	117.513	2.662
7		1200	60		141.52	19.325	7.901
8		1200	120		99.08	38.755	5.014
9		1200	180		71.11	199.510	2.634

$$MRR = \frac{I_W - F_W}{\rho \times C_T} \qquad \bullet \qquad (3.2)$$

$$PE = V \times I \times PT \qquad (3.3)$$

where X is a load applied in N, I_m and F_m denotes the initial and final masses in mm^3/g of HFRP composite; ρ is the density of the HFRP composite; PT is cutting/process time in s, V and I represent voltage and current used during each experiment, respectively. In analysis, MRR is taken as higher-the-better and process time, thrust force and process energy are lower-the-better.

3.3 Results and Discussion

3.3.1 Effect of Process Parameters on Thrust Force

The effect of process parameters such as spindle speed and feed rate on thrust force is depicted in Fig. 3.4a, b. It is clearly noticed that input parameters like SS and FR are more influenced parameters on thrust force. It is noticed from Fig. 3.4a that the value of independent parameter spindle speed increased from 500 to 800 rpm, and the value of dependent parameter thrust force is a drastic change from 116.20 to 108.33 N. Further, the response parameter thrust force is gradually decreased 108.33–103.90 N with increasing the value of the input parameter spindle speed from 800 to 1200 rpm. This is because the amount of heat generated increases with spindle speed [21, 25]. This results in the thrust force which is reduced. However, the influence of independent parameter feed rate on response parameter thrust force as shown in Fig. 3.4b that the process parameter feed rate increases from 60 to 120 mm/rev, and the value of response parameter thrust force is decreased from 148.22 to 102.87 N, further it is decreased from 102.87 to 75.69 N. This is because heat dissipation is taking place through the chip, resulting in lower thrust force which is achieved [17].

3.3.2 Effect of Process Parameter on MRR

The impact of input parameters such as spindle speed and feed rate on MRR is shown in Fig. 3.5a, b. It is observed from Fig. 3.5a that the input parameter SS increased from 500 to 800 rpm, the value of output parameter MRR significantly increases from 40.29 mm^3/rev to 69.77 mm^3/s, and further, it is increased from 69.77 to 85.86 mm^3/s with parameter spindle speed gradually increases from 800 to 1200 rpm. This is because the processing time decreases with an increase in cutting forces, resulting in higher MRR which is achieved. Moreover the process parameter feed rate is positively influenced by response parameter MRR as shown in Fig. 3.5b.

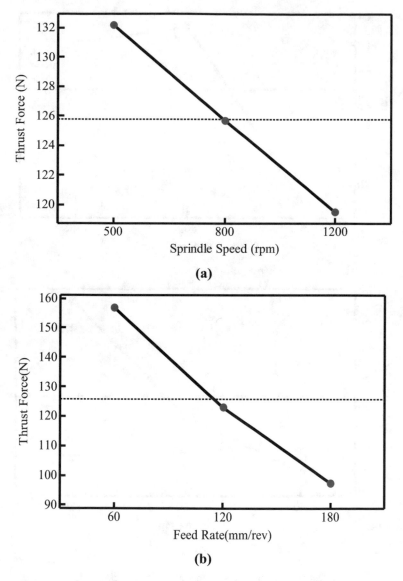

Fig. 3.4 a, b Main effect plot for thrust force versus input parameters levels

The independent parameter feed rate varies from 60 to 80 mm/rev, the response parameter MRR significantly increases from 27.89 to 38.23 mm^3/s, and further, it is increased from 38.23 to 124.34 mm^3/s. This is because tool penetration into the work material and cutting forces increases, resulting in higher MRR which is achieved [21, 25].

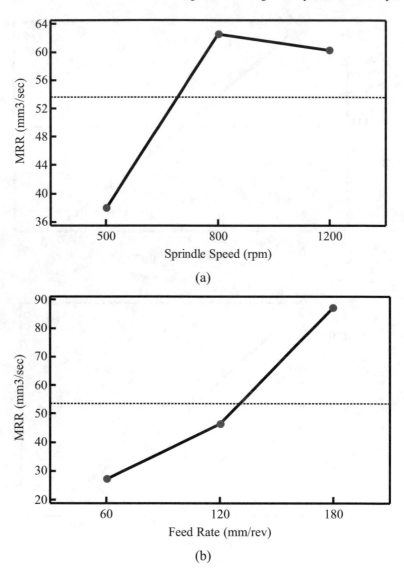

Fig. 3.5 a, b Main effect plot for MRR versus input parameters levels

3.3.3   Effect of Process Parameter on Process Energy

The impact of input parameters such as spindle speed and feed rate on process energy of HFRP composite analyzed as shown in Fig. 3.6a, b.

It is clearly identified from Fig. 3.6a, b that parameter feed rate's negative effect on process energy and parameter spindle speed shows the minimal effect on process energy. It is also identified from Fig. 3.6b that the independent parameter feed rate

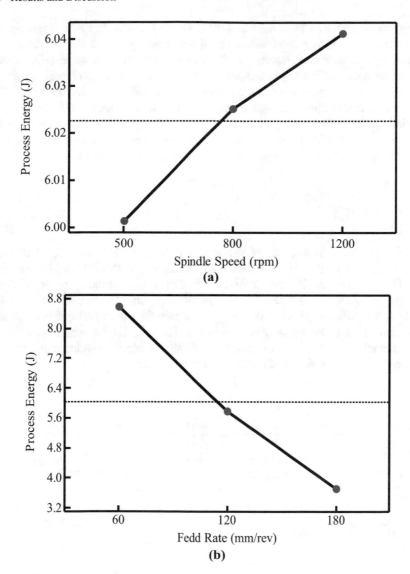

Fig. 3.6 a, b Main effect plots for process energy versus input parameters levels

increases from 60 to 120 mm/rev, and the response parameter process energy gradually decreases from 7.86 to 4.58 J. Further, the dependent parameter process energy
marginal decrement from 4.62 to 2.69 J with process parameter feed rate increases
from 120 to 180 mm/rev. This is because the cutting forces on work material are
decreased and process time is increased, resulting in lower process energy. Also,
observed from Fig. 3.6a that, for 5% HFRP fiber composition graph shows, the input
parameter spindle speed increases from 500 to 800 rpm, and the out parameter process

energy value decreases from 4.93 to 4.90 J. This is because, in this range of speed, the machining of HFRP composite, cutting forces acting on the work material, and process time is decreased, resulting in the value of process energy which is reduced. But the process parameter spindle speed increases from 800 to 1200 rpm, the dependent parameter process energy value significantly increases from 4.90 to 5.18 J. This is because the tool moves at high speed and improper contact between work material and tool and takes more process time for machining of HFRP composite. In the case of feed rate (Fig. 3.6b), the response parameter process energy is drastically decreasing from 8.2 to 3.8 J when the feed rate varies from 60 to 180 mm/rev.

3.3.4 ANOVA

ANOVA is termed as analysis of variance is used to analyze the impact of input parameters on HFRP composite during the machining with CNC drilling process. MINITAB 19 software is used for ANOVA and the results of the same are tabulated in Tables 3.3, 3.4, 3.5, 3.6 and 3.7. In ANOVA, by using the stepwise elimination method, insignificant parameters are removed to adjust the fitted quadratic model [26]. If the probability value (P-value) is less than 0.05 (alpha value), which indicates that all the parameters are significant. Similarly, higher F-value represents the performance characteristics of the input parameters [27].

Table 3.3 ANOVA for thrust force

Source	DF	Adj SS	Adj MS	F-value	P-value
Regression	2	7953.1	3976.57	148.63	0
SS	1	219.6	219.56	8.21	0.029
FR	1	7733.6	7733.58	289.06	0
Error	6	160.5	26.75		
Total	8	8113.7			

$R^2 = 98.03\%$, R^2 (adj) $= 96.84\%$, R^2 (Pred) $= 90.88\%$

Table 3.4 ANOVA for MRR

Source	DF	Adj SS	Adj MS	F-value	P-value
Regression	3	22596	7531.9	9.39	0.017
SS	1	2738	2737.9	3.41	0.024
FR	1	1362	1361.9	1.7	0.249
SS × FR	1	6216	6216.2	7.75	0.139
Error	5	4012	802.4		
Total	8	26,608			

$R^2 = 95.77\%$, R^2 (adj) $= 93.23\%$, R^2 (Pred) $= 82.37\%$

Table 3.5 ANOVA for process energy

Source	DF	Adj SS	Adj MS	F-value	P-value
Regression	1	37.6251	37.6251	487.22	0
FR	1	37.6251	37.6251	487.22	0
Error	7	0.5406	0.0772		
Total	8	38.1657			

$R^2 = 98.58\%$ R^2 (adj) $= 98.38\%$ R^2 (Pred) $= 97.86\%$

Table 3.6 Normalized (N_{ij}) values of CNC drilling parameters

Exp. No.	Normalized (N_{ij}) values		
	N_{ij} of (TF)	N_{ij} of (MRR)	N_{ij} of (PE)
1	0.2028	0.0147	0.2125
2	0.1053	0.0192	0.0843
3	0.0618	0.0466	0.0257
4	0.1922	0.0227	0.2154
5	0.0909	0.0438	0.0770
6	0.0460	0.2123	0.0269
7	0.1727	0.0057	0.2366
8	0.0846	0.0231	0.0953
9	0.0436	0.6119	0.0263

Table 3.7 Ranking of assessment values (y_i) values for CNC drilling process

Exp. No.	% Composition	Input parameters				
		SS (rpm)	FR (mm/rev)	Dielectric fluid	y_i values	Rank
1	5%	500	60	Tap water	0.611	9
2		500	120		0.747	7
3		500	180		1.021	4
4		800	60		0.690	8
5		800	120		0.919	5
6		800	180		1.106	2
7		1200	60		0.866	6
8		1200	120		1.069	3
9		**1200**	**180**		**1.256**	**1**

It has been observed from the results (Table 3.4) that both feed rate and spindle speed is more significant factor on experimental results. First, most significant parameter for thrust force is tested using F-test. F-value for feed rate (linear parameter) yields the highest with 289.06 and P-value is exactly 0.000. It implies that the process

parameter feed rate is the most significant in determining the performance characteristic thrust force for 5% HFRP composite machining while other parameters are
found least significant. Furthermore, P-test (i.e., statistical significance) followed by
coefficient of determination (R^2) and adjusted-coefficient of determination (Adj-R^2)
is done for each of the responses. Since the value of p for each of the responses
obtained as less than 0.05 and 98.03% as R^2 and 96.84 as Adj-R^2, this indicates that
process and associated parameters are statistically significance and better fitness of
the data.

Table 3.5 shows the ANOVA result for MRR, it is observed that the F-value
of spindle speed (linear parameter) is larger at 3.41 and P-value is exactly 0.019,
which is less than 0.05. It implies that the spindle speed is most significant for MRR.
While the R^2 obtained for MRR is 95.77% and the adjusted R^2 is 93.23%, shows the
presented model fits the data well enough. Although an interaction parameter spindle
speed × feed rate exists, it is not so significant because of lesser values of F- and
P-values. Similarly, the result of ANOVA for process energy of HFRP composites is
tabulated in Table 3.7. The results of ANOVA show that the F-value of feed rate is
larger at 487.22 and P-value is exactly $0.0 < 0.05$. It means that process parameter,
i.e., feed rate is most significant for process energy. While the R^2 obtained for process
energy is 98.58% and the adjusted R^2 is 98.38%, the presented model fits the data
well.

3.3.5 Empirical Model

In this section, empirical models development has been discussed. Using multiple
regression analysis in Minitab software, individual empirical models have been
generated and plotted. Equations (3.4–3.6) contain input variables (spindle speed,
feed rate) on the right side and output variable on the left side [28].

$$TF = 194.4 - 0.0157 \times SS - 0.588 \times FR - 0.000013(SS \times FR) \qquad (3.4)$$

$$MRR = 13 - 0.039 \times SS + 4.86 \times FR + 0.0009(SS \times FR) \qquad (3.5)$$

$$PE = 10.10 - 0.00084 \times SS - 0.0389 \times FR - 0.000011(SS \times FR) \qquad (3.6)$$

The empirical models (Eqs. 3.4–3.6) will be used for prediction of the response
parameters, i.e., thrust force, MRR, and process energy for the different input conditions of the CNC drilling process on machining of HFRP composites. Further, the
normality of residuals are plotted to determine the normality of the data points of
thrust force MRR, and process energy using normality plots, and their graphical representation is shown in Fig. 3.7a–c. The graph shows that all points of the responses
for thrust force, MRR, and process energy are nearer to the linear line and fallows
normally distributed with 95% CI.

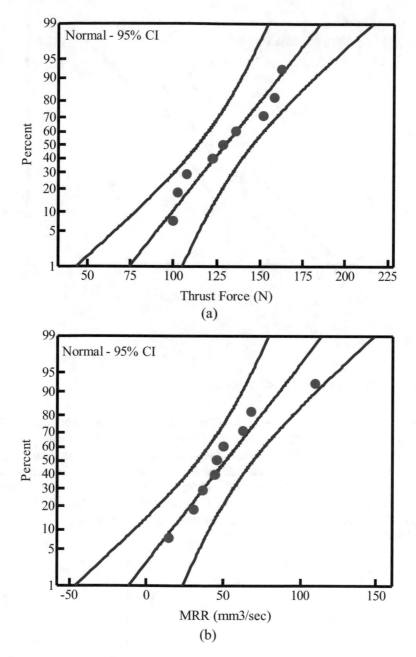

Fig. 3.7 Probability plot for: **a** Thrust force, **b** MRR, **c**, **d** Process energy of HFRP composites

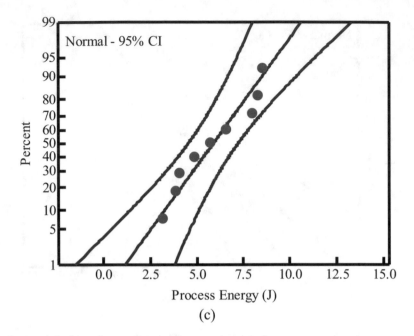

Fig. 3.7 (continued)

3.3.6 Modeling and Optimization

Optimization of input parameters of the CNC drilling process on machining of HFRP composite (5, 7.5, and 10%) is performed using a multi-objective optimization ratio analysis (MOORA) method [24–26]. The parameters process time, thrust force, MRR, and process energy are considered as response parameters for each of the HFRP composite, while spindle speed and feed rate as process parameters. In this method first, designing a decision matrix is carried out based on Taguchi (L_9) using Eq. (3.7). The output results of the decision matrix for each of the HFRP composites are tabulated in Table 3.2.

$$D_{ij} = \begin{bmatrix} & C_1 & C_2 & \dots & C_n \\ A_1 & Y_{11} & Y_{12} & \dots & Y_{1n} \\ A_2 & Y_{21} & Y_{22} & \dots & Y_{2n} \\ \dots & \dots & \dots & \dots & \dots \\ A_m & Y_{m1} & Y_{m2} & \dots & Y_{mn} \end{bmatrix} \tag{3.7}$$

where D_{ij} is the decision matrix contains response values of ith alternatives on jth criterion, i.e., $Y_{11}, Y_{12}, \dots, Y_{mn}$; $C_1, C_2, \dots, C_n$ represents the number of criteria or no. of response parameters; $A_1, A_2, \dots$, is a no. of alternatives or experiments.

Table 3.8 Confirmatory tests for CNC drilling process of HFRP composites

Input conditions	Output parameters	Experimental results	Confirmatory tests results
Exp. No. 9 SS (1200 rpm) and FR (180 mm/rev)	TF (N)	71.11	70.22
	MRR (mm³/s)	199.51	198.63
	PE (J)	2.634	2.534

Next, the conversion of different units of CNC response datas into a comparable sequence data is done by the normalization process using Eq. (3.8). This step is essential because processing of HFRP composites in CNC process problem contains different units of output parameters like thrust force, MRR, and process energy. In order to evaluate the overall assessment for CNC process, one needs to convert these output parameters into comparable sequence. The result of normalization matrix for CNC drilling process on HFRP composites is tabulated in Table 3.6.

$$N_{ij} = \frac{D_{ij}}{\left[\sum_{i=1}^{m} Y_{ij}^2\right]^{1/2}} \quad \text{where } j = 1, 2, \ldots, n \tag{3.8}$$

After that, the determination of overall assessment values (y_j) for each one of the alternatives or exp. no is done using Eq. (3.9). This step converts the entire multi-response optimization into a single optimization problem. Later, ranking of the exp. no or alternatives are done based on the assessment values [27–31]. The choice with greater assessment values (y_j) gives optimal operating conditions to get higher performance for the system compared to the other experiment number. The outcomes of the ranking are tabulated in Table 3.8.

$$y_j = \sum_{i=1}^{g} N_{ij} - \sum_{i=g+1}^{n} N_{ij} \tag{3.9}$$

where N_{ij} denotes normalized performance values of ith output parameters, g denotes the no. of parameters to be maximized, $(n - g)$ indicates the no. of parameters to be minimized, y_j indicates assessment values of ith parameters w.r.t the all jth exp. runs.

The result shows that, for 5% HFRP composites in the CNC drilling process, Exp. No. 9 achieved the highest attainment values among the other runs. The assessment (y_j) value for optimal setting obtained is 1.256. This shows the highest assessment (y_j) value among the other experiments. The following optimal setting yields the best combination of process variables are spindle speed (1200 rpm, level 3) and feed rate (180 mm/rev, level 3). The optimal setting obtained using MOORA method provides the most optimal values for CNC drilling process which have less influence on the performance of CNC drilling process during the machining of HFRP composites. Also optimal setting creates less environmental impacts (because lower process energy, lower thrust process and higher MRR as well as produce less environmental pollution due to the use of tap water) during the machining of HFRP composites. Also it improves the quality of the product, reduces the final product costs, and improves the machining efficiency of the CNC drilling process.

Therefore, the optimal values such as MRR, thrust force and process energy and tap water as dielectric fluid results in the CNC drilling process toward the green manufacturing process. Hence, it is recommended that optimal setting, i.e., spindle speed (1200 rpm, level 3), and feed rate (180 mm/rev, level 3), is used in the CNC drilling for effective machining of HFRP composites under tap water. Furthermore, the work suggested that the proposed method, i.e., MOORA, can be utilized for finding the best possible combination of input parameters for CNC drilling process in green manufacturing environment and other processes as well.

3.3.7   Confirmatory Analysis

Further, confirmatory tests are performed to verify the results obtained via. MOORA method. The optimal setting, i.e., spindle speed (1200 rpm, level 3) and feed rate (180 mm/rev, level 3) are used for confirmatory experiments and the corresponding results are tabulated in Table 3.8. The results show that confirmatory test results are comparable and acceptable with experimental results for the optimal setting.

3.4   Summary

The manufacturing and processing of HFRP composites in the CNC drilling process is discussed in this chapter. Manufacturing of 5% hemp fiber composition with 8 mm thickness is prepared by conventional hand layup process while processing of HFRP composites is done by CNC drilling process in green machining environment. Work considered two (feed rate and spindle speed) parameters with dielectric fluid (tap water) as constant and they are varied as per the Taguchi (L_9) design. The main reason for using tap water is for the ecological issues because tap water is good working environment and does not produce any kind of harmful gases during the machining in comparison with other fluids. This results in less environmental and health issues and makes the CNC drilling process toward the green manufacturing. The results show that parameter spindle speed and feed rate are observed to be more significant parameters on thrust force, MRR and process energy. The optimal parameters obtained at spindle speed(1200 rpm, level 3) and feed rate(180 mm/rev, level 3) which provides as thrust force (71.11 N), MRR (199.51 mm^3/s), and process energy (2.634 J). This optimal parameter have less influence on responses of CNC drilling process and play an important role in the reduction of environmental impact (because lower process energy, lower thrust process, and higher MRR as well as produce less environmental pollution due to the use of tap water) during the processing of HFRP composites in CNC drilling. Additionally, ANOVA and empirical models for thrust force, MRR, and process energy of HFRP composites are performed and found that they are statistically significant and fit the data well enough with the experiment results with the 95% confidence level. Normality of the data points is analyzed by plotting the probability plots and plots show that data are normally distributed and normality assumption is valid. Last, confirmatory tests are performed to verify the results obtained via. MOORA method. The results show that confirmatory test results are comparable and acceptable with experimental results for the optimal setting.

From the above observations, the HFRP composite gives better execution in the CNC drilling process. The parametric settings got from the MOORA technique can be utilized as the ideal setting for the CNC drilling during the machining of HFRP composites. Hence, it is recommended that optimal setting, i.e., spindle speed (1200 rpm, level 3), and feed rate (180 mm/rev, level 3), is used in the CNC drilling for effective machining of HFRP composites under tap water. Furthermore, the work suggested that the proposed method, i.e., MOORA, can be utilized for finding the best possible combination of input parameters for CNC drilling process in green manufacturing environment and other processes as well.

References

1. M. Karus, D. Vogt, European hemp industry: cultivation, processing and product lines. Euphytica **140**, 7–12 (2004)
2. L.Y. Mwaikambo, M.P. Ansell, Chemical modification of hemp, sisal, jute, and kapok fibers by alkalization. J. Appl. Poly. Sci. **84**, 2222–2234 (2016)
3. M. Kostic, B. Pejic, P. Skundric, Quality of chemically modified hemp fibers. Bioresour. Technol. **99**(1), 94–99 (2008)
4. S. Suhara, S. Mohini, Water absorption properties of short hemp—glass fiber hybrid polypropylene composites. J. Compo. Mater. **41**, 1871 (2007)
5. A. Thygesen, *Properties of Hemp Fibre Polymer Composites*, vol. 11 (Denmark, 2006). ISBN 87-550-3440-3
6. N.S.V. Gupta, K.V.S. Rao, An experimental study on sisal/ hemp fiber reinforced hybrid composites. Mater. Today: Proc. **5**(2), 7383–7387 (2018)
7. S.M.M. Somashekar, Investigation on mechanical properties of hemp-E glass fiber. Int. J. Mech. Eng. Technol. **7**(3), 182–192 (2016)
8. J. Naveen, J.M. Amuthakannan, M. Chandrasekar, Mechanical and physical properties of sisal and hybrid sisal fiber-reinforced polymer composites, in *Mechanical and Physical Testing of Biocomposites, Fibre-Reinforced Composites and Hybrid Composites* (Woodhead Publisher, 2018), p. 427–439
9. G. Sèbe, N.S. Cetin, C.A.S. Hill, M. Hughes, RTM hemp fibre-reinforced polyester composites. Appl. Compos. Mater. **7**, 341–350 (2000)
10. G.D. Babu, K.S. Babu, B.U.M. Gowd, Optimization of machining parameters in drilling hemp fiber reinforced composites to maximize the tensile strength using design experiments. Indian J. Eng. Mater. Sci. **20**, 385–390 (2013)
11. K. Patel, Investigation on drilling of banana fibre reinforced composites, in *2nd International Conference on Civil, Materials and Environmental Sciences*, p. 201–205 (2015)
12. V.A. Prabu, S.T. Kumaran, M. Uthayakumar, V.A. Prabu, S.T. Kumaran, M. Uthayakumar, Performance evaluation of abrasive water jet machining on banana fiber reinforced polyester composite. J. Nat. Fibers **00**(00), 1–8 (2016)

13. R. Selvam, L. Karunamoorthy, N. Arunkumar, Investigation on performance of abrasive water jet in machining hybrid composites. Mater. Manuf. Proc. **32**(6), 1–28 (2016)
14. Jagadish, A. Ray, Optimization of process parameters of green electrical dichrage machining using PCA. Int. J. Adv. Manuf. Technol. **87**, 1299–1311 (2015)
15. S. Jayabal, U. Natarajan, Drilling analysis of coir-fibre-reinforced polyester composites. Bull. Mater. Sci. **34**(7), 1563–1567 (2011)
16. U. Patel, K. Patel, P. Gohil, V. Chaudhary, Investigations on drilling of unidirectional hemp-polyester composites. Int. J. Mech. Eng. Technol. **10**(1), 707–718 (2019)
17. H.R. Maleki, M. Hamedi, M. Kubouchi, Y. Arao, Experimental study on drilling of jute fiber reinforced polymer composites. J. Compo. Mater. **0**(0), 1–13 (2018)
18. V.V. Kumar, A. Shirisha, T. Praveen, Optimization of process parameters in drilling of natural fibre composite. Materials 12–15 (2014)
19. M. Ramesh, K. Palanikumar, K.H. Reddy, Experimental Investigation and Analysis of Machining Characteristics in Drilling Hybrid Glass-Sisal-Jute Fiber Reinforced Polymer Composites, in *5th International and 26th All India Manufacturing Technology, Design and Research Conference*, vol. 461, p. 1–6 (2014)
20. R. Manickam, A. Gopinath, Measurement and analysis of thrust force in drilling sisal-glass fiber reinforced polymer composites. IOP Conf. Ser.: Mater. Sci. Eng. **197**(1), 1–7 (2017)
21. D. Chandramohan, K. Marimuthu, Drilling of natural fiber particle reinforced polymer composite material. Int. J. Adv. Eng. Res. Stud. **1**(1), 134–145 (2011)
22. V.A. Prabu, S.T. Kumaran, M. Uthayakumar, Performance evaluation of abrasive water jet machining on banana fiber reinforced polyester composite. J. Nat. Fibers **14**, 450–457 (2017)
23. R. Selvam, L. Karunamoorthy, N. Arunkumar, Investigation on performance of abrasive water jet in machining hybrid composites. Mater. Manuf. Process **17**(32), 700–766 (2017)
24. Rajakumaran M. Jagadish, A. Ray, Investigation on mechanical properties of pineapple leaf- based short fiber-reinforced polymer composite from selected Indian (northeastern part) cultivars. J. Thermo Compo. Mater. **1**(1), 1–19 (2018)
25. S. Jayabal, U. Natarajan, Optimization of thrust force, torque, and tool wear in drilling of coir fiber-reinforced composites using Nelder-Mead and genetic algorithm methods. Int. J. Adv. Manuf. Technol. **51**, 371–372 (2010)
26. S. Jayabal, U. Natarajan, Drilling analysis of coir-fibre-reinforced polyester composites. **34**(7), 1563–1567 (2011b)
27. J. Kechagias, G. Petropoulos, Application of Taguchi design for quality characterization of abrasive water jet machining of TRIP sheet steels. Int. J. Adv. Manuf. Technol. **62**, 635–643 (2012)
28. M.K. Pradhan, Estimating the effect of process parameters on MRR, TWR and radial overcut of EDMed AISI D2 tool steel by RSM and GRA coupled with PCA. Int. J. Adv. Manuf. Technol. **68**, 591–605 (2013)
29. R.V. Rao, V.D. Kalyankar, Optimization of modern machining processes using advanced optimization techniques: a review. Int. J. Adv. Manuf. Technol. **73**(5–8), 1159–1188 (2014)
30. R. Kumar, K. Kumar, S. Bhowmik, Optimization of mechanical properties of epoxy based wood dust reinforced green composite using taguchi method. Int. Conf. Adv. Manuf. Mater. Eng. **5**, 688–696 (2014)
31. S. Bhowmik, Jagadish, K, Gupta, Modeling and optimization of advanced manufacturing processes, in *Manufacturing and Surface Engineering*, 1st edn (Springer International Publishing, 2019), p, 1–74. https://doi.org/10.1007/978-3-030-00036-3

Chapter 4
Manufacturing and Processing of Banana Fiber-Based Polymer Composite

Abstract Recently, many industries try to minimize petroleum-based materials and products due to the ecological issues and the requirement for more flexible materials. This makes industries to use polymer composites based on natural fibers due to their unique properties. In fact, use of banana fiber in manufacturing of polymer composites is increased recently for various applications. Apart from the physical, chemical and mechanical properties, these composites undergo various secondary operations before actual use. In this chapter, an attempt is made on manufacturing and machinability study of short banana fiber-reinforced polymer (BFRP) composites. Convectional hand layup process is used for manufacturing while CNC milling process is used for machinability of BFRP composite. Processing parameters like % fiber, speed and feed rate on MRR, process time, and thrust force are studied according to the Taguchi (L_9). Using AHP-TOPSIS method, optimal processing parameters for CNC milling are established. Results show that hand layup process and CNC milling process is feasible in processing of BFRP composites and produce better products. Ideal setting for BFRP composites obtained as % fiber: 10%, speed: 500 rpm and feed rate: 180 mm/rev and parameter %fiber and speed found more significance during the processing of BFRP composite.

Keywords BFRP composite · CNC milling process · Parametric analysis · ANOVA · Regression analysis · Optimization

4.1 Introduction

Banana plants are the oldest and most economical farming crop which belongs to the *Musaceae* family. Globally, India ranks first in banana production by producing around 14 million metric tonnes per year. Banana plants (*Musa sapientum*) are abundantly available in the Chhattisgarh and coastal regions in India. Mostly, in India banana is cultivated, viz. Maharashtra, Kerala, Andhra Pradesh, Orissa, Tamil Nadu, West Bengal, Assam, Bihar, and Karnataka. Globally, there are many varieties of bananas available, and dessert-type and culinary-type bananas are two most commercial varieties of *Musa* family [1]. One of the most advantages in the banana plant

© The Author(s), under exclusive license to Springer Nature Switzerland AG 2021   59
Jagadish and S. Bhowmik, *Manufacturing and Processing of Natural Filler Based Polymer Composites*, SpringerBriefs in Applied Sciences and Technology,
https://doi.org/10.1007/978-3-030-65362-0_4

(Fig. 4.1) is every part of the plants which are useful like pseudostem, midrib, banana leaf, leaf sheath, inflorescence (flower), and banana fruit, out of which pseudostem and banana fruit are a most beneficial part of the banana plant. The fibril extracted from the pseudostem of the banana plant can be utilized in many applications like greeting cards, lamp stands, pen stands, decorative papers, rope, paper, filter paper, textiles, shoes, furniture, bags .baskets, rugs, and mats. Now day's banana fibers are widely used in automobile industries and structural applications because of superior flexural strength, impact strength, eco-friendly, and biodegradability and has good mechanical properties [2, 3]. The banana plant generally, green, light yellow and creamy color and fiber is grown approximately in-between 3 m (10 ft) and 12.2 m (40 ft), long surrounding with 8–12 large leaves up to 2.7 m (9 ft) long and 0.61 m wide (2 ft), based on the method of extraction of the fibrils from the stem. The chemical composition of banana fibers consists of cellulose (48–60%), hemicelluloses (12–16%), lignin (15–22%), and pectin (3–5%) and fat and wax (1–2%). Because of the high content of cellulose, degree of crystallinity, banana fibrils are selected as reinforcement material in the present work.

In addition, banana fibers are abundantly available with less in density and economically cheap and have larger industrial applications as compared to other natural fibers (jute, coir, flax, coconut, sisal, hemp, bamboo, bagasses, rice husk, etc.). Meanwhile, some extent of work has been done on the banana fiber-based composites with their physical and mechanical properties. Kalia et al. [4], Prasad et al. [5], Mishra et al. [6] and Mejia et al. [7] presents chemical treatment on banana fibers with NaOH solution. The results show that one percent of the NaOH solution for banana fiber gives good crystallinity index, thermal stability, and better

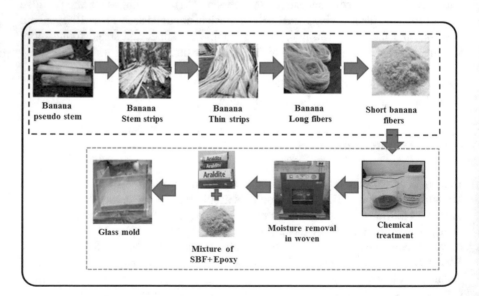

Fig. 4.1 Extraction + fabrication steps of BFRP composite

mechanical properties (i.e., flexural strength, impact strength, and tensile strength). Alavudeen et al. [8] studied the mechanical properties of the banana/kenaf fiber-reinforced hybrid polyester composites and its effect on random and woven fabric orientation patterns. The result shows that the minimum stress is induced at the composite interface due to the load distribution in the fiber direction. Zin et al. [9] studied the effect of various fiber loadings on the flexural and thermal properties of bananas. In the study polymers are used to manufacture filler in the form of fibers or particles to produce products with the necessary thermal, mechanical, and electrical characteristics. They suggested that the initial degradation temperature for both banana glass and pineapple leaf fiber—glass hybrid composites should be given after TGA studies. Mohan and Kanny [10] studied nano-clay infused banana fiber and its effects on mechanical properties. The results reveal that infused nano-clay into the banana fibers the tensile, interfacial, and thermal properties are improved.

Despite mechanical and chemical characterization, processing or machining of banana fiber-reinforced polymer (BFRP) composites is also important in manufacturing the engineering products. Also, these composites undergo various secondary operations apart from the primary operations before actual use. Commonly drilling, milling, turning, and cutting are performed to get the actual product. Hence, machining or cutting with different shapes and higher productivity at minimum machining costs are essential to study the machinability of BFRP composites. The machining of BFRP composites is possible using conventional and non-conventional machining processes. Some of the machining study for other natural fibers-based polymer composites is that Jayabal and Natarajan [11] studied the effect of drilling parameters (drill diameter, feed and speed on torque, thrust force and tool wear) on the coconut polyester composite. The work concluded that with increased tool wear, tool life becomes negligible. Jayabal and Natarajan [12] investigated the machining properties of coir-polyester-based polymer composites using traditional drilling process. Three independent parameters, namely drill diameter, feed rate and speed, were varied according to the Taguchi (L_9) method and corresponding response torque, thrust force and tool wear were examined. Work reported that optimum values gave greater reduction of tool wear in machining of coir fiber-based polymer composites. Athijayamani et al. [13] performed the machinability of roselle-sisal hybrid composites using drilling process. Speed, feed and tool diameter as input variables and torque and thrust force as output variables were considered for the study. Work also employed ANN and regression model for the prediction of optimal values for drilling process. Investigation results shows that ANN gave better performance in prediction compared to the regression model. Chandramohan and Marimuthu [14] performed the machinability study of sisal and roselle-bio-epoxy resin composite materials in drilling process. Work suggested that larger diameter drills and higher feed rates gave optimum results for sisal and roselle-bio-epoxy resin composite materials. They also observed that volume fraction of fiber and higher feed increases the thrust force but decreases with increase of increase of cutting speed. The machining creates various machining issues due to their intermolecular structure, delamination problem, fiber pull-out during processing, incorrect machining, lower productivity, etc. [15–17]. Further, both conventional and non-conventional machining processes generate various means of wastages, resulting in serious ecological and health-related concerns. To overcome these issues, the present chapter attempted to use CNC milling process under dry machining conditions for processing/machining of BFRP composites.

Moreover, the processing/machining of BFRP composites using the CNC milling process under the green machining environment has not been done so far in the literature. Here, first, BFRP composites are fabricated in three varieties of compositions like (10, 15, and 20%) using a well-known hand layup process. Thereafter, an experimental investigation is carried out based on Taguchi (L_9-OA) to analyze the green processing attributes (such as MRR, process time, and thrust force) to the variation in process parameters like % fiber, speed and feed rate. Further, parametric, ANOVA, and multi-regression analysis are performed for suitability and statistical significance of the CNC milling process on BFRP composites. Subsequently, multi-response optimization is done using an integrated method AHP-TOPSIS method. Here, the AHP method is employed for the extraction of priority weights of the response parameters while the TOPSIS method for ranking of the experimental runs. However, no work is available in the literature on the feasibility of using AHP-TOPSIS optimization of CNC milling parameters processing of BFRP composites. At last, results obtained via the AHP-TOPSIS method are verified using a confirmatory analysis.

4.2   Materials Manufacturing and Machining Details

4.2.1   Manufacturing of BFRP Composites Using Hand Layup Process

In this study, short banana fibers are extracted from the banana plant used for the preparation of BFRP composites. The banana plant is taken from the local vendor. Firstly, the pseudo-banana stem is obtained from banana plants and pseudostem consists of a number of round layers up to the core of the banana plant (Fig. 4.1). The layers of pseudo-banana stem are removed one by one till white pulp of pseudostem arrives at the core of banana plant. The width of the pseudo-banana stem layer is 15–20 cm (approx.) and the length of stem depends on the height of the banana plant and then these stems are made into a thin strip of size 10–20 cm. These thin strips are exposed in the sunshine for 48 h to make sure the strips are free from moisture content [18]. The removal of moisture content is an extreme priority because it will affect the mechanical properties of the composite. The long thin strips are processed into very fine and long fibers and then long fibers are made into short fibers of size 1–2 mm using the milling machine.

Once completed the extraction process, short banana fibers are undergone various chemical treatments with NaOH solution, alcohol, ethanol solution to remove the impurities from the fibers followed by moisture removal in woven [18]. Later, fibers are mixed with epoxy resin (Araldite LY 556, Hardener HY 951) in the ratio of 10:8

by weight. During the mixer, homogenous of the resin and fiber is maintained by proper stirring using a mechanical stirrer. The composite with 10, 15 and 20% fiber composition are prepaid. The mixer in the glass mold kept for 24–30 h under room temperature for proper cure. Finally, specimens are removed from the mold and cut as per the required shape and size for further use.

4.2.2 Machining Details

The CNC milling machine is used for experimental work with a drill bit instead of milling tools as shown in Fig. 4.2. The various specifications of the CNC machine are given in Table 4.1.

During the experimentation three independent parameters or control factors like % of fiber, speed (rpm), feed rate (mm/s)with three levels (Table 4.2) while thrust force (TF) in N, process time (PT) in sec, and MRR in cc/s as output parameters. Slot cutting tests are conducted on BFRP composite as per the Taguchi (L_9-OA)

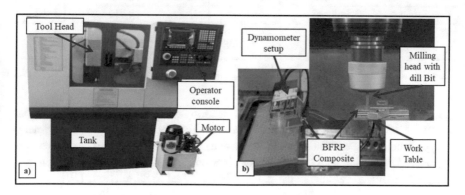

Fig. 4.2 a CNC milling machine, **b** tool head setup

Table 4.1 CNC milling machine specifications

Manufacturer	MTAB Pvt. Ltd.
Size	275 × 150 × 250
Model	Trainer CNC milling
CNC controller	Sinumerik 808D
Axis motor and drive	AC Servo (1.3 Nm)
Spindle motor	1 HP with VFD (3PH)
Spindle motor speed	100–3000 rpm
Voltage	230 V
Amperes	2.5 A

Table 4.2 CNC Milling process parameters

Input parameters	Units	Symbol	Levels		
			1	2	3
Composition	%	A	10	15	20
Cutting speed	rpm	B	500	800	1200
Feed rate	mm/s	C	60	120	180

Table 4.3 Experimental results of CNC milling

Exp. No.	Input parameters			Output parameters		
	% of fiber (A)	Speed (B) (Rpm)	Feed rate (C) in mm/s	Thrust force (TF) in N	Process time (PT) in s	MRR in mm^3/s
1	10	500	60	22.797	4.210	0.391
2	15	800	120	12.625	2.460	0.198
3	20	1200	180	19.415	3.790	0.412
4	10	500	120	20.136	2.970	0.312
5	15	800	180	11.682	2.060	0.194
6	20	1200	60	22.327	4.810	0.402
7	10	500	180	14.302	3.650	0.373
8	15	800	60	18.345	2.130	0.154
9	20	1200	120	20.149	5.320	0.362

[19, 20] using Ø6 mm drill bit without lubrication. Total nine slots with dimensions 12 mm × 12 mm are cut on BFRP composites. Total of three trials of experiments are performed for each setting and their average values of the output parameters are considered for further analysis (Table 4.3). The output parameters thrust force, MRR, and process time is calculated using the following empirical relations [19].

$$TF = \frac{X \times 9.81}{1000} \tag{4.1}$$

$$MRR = \frac{I_W - F_W}{\rho \times C_T} \tag{4.2}$$

where X is load applied in N, I_w and F_w denote initial and final masses in mm^3/g of BFRP composite, and ρ is density of the BFRP composite. The parameter MRR as larger is better while process time and thrust force as smaller is the better is considered in analysis.

4.3 Results and Discussion

4.3.1 Effects of Input Parameters on Thrust Force

The influence of process parameters such as A, B, and C on thrust force is depicted in Fig. 4.3. It is observed that thrust force increases from 14 to 21 N with an increase in % fiber (A) from 10 to 30%. This is because higher fiber concentration in the matrix materials offers more resistance to material removal by the tool resulting in the requirement of more force to cut the materials (Fig. 4.3a). The parameter B (i.e., speed) on TF shows decrement manner from 20.63 to 14.22 N when speed (B) is increased from 500 to 1200 rpm. This is because an increase in speed, the amount of heat generated also increases. Then the generated heat gets accumulated around the wall of the machined surface due to the lower thermal conductivity which leads to the plastic deformation of BFRP composites, resulting in the reduction of thrust force (Fig. 4.3b). However, the influence of feed rate, i.e., C on thrust force also shows a similar decrement manner from 21.156 to 17.637 N when C is increasing from 60 to 120 mm/rev. Further, it is decreased from 17.637 to 15.133 N when the C is increased from 120 to 180 mm/rev because heat dissipation is taking place through the chip, resulting in (Fig. 4.3c) the lower thrust force which is achieved [19, 20].

4.3.2 Effects of Input Parameters on Process Time

The impact of input parameters such as A, B, and C on process time is discussed and the main effect plot of A, B and C on process time is shown in Fig. 4.4. The main effect plots show that response process time found decrement with an increase in % of fiber increment. This is because higher % fiber concentration increases the strength and hardness of BFRP composite and allows the BFRP composite to bear more resistance in machining, resulting in lesser thrust force (Fig. 4.4a). The parameter B (i.e., speed) also shows a similar decrement (from 3.610 to 2.217 s) with speed increased from 500 to 800 rpm (Fig. 4.4b). This is due to the fact that, at medium speed, there is effective contact between tool and workpiece, and also molecular structure is poor due to less % fiber composition [20, 21]. But when the speed increased from 800 to 1200 rpm, the process time gets increased from 2.217 to 3.167 s, because the drill bit rotates at higher speed and improper contact between work material and tool. However, the influence of C (i.e., feed rate) on process time (Fig. 4.5) shows, increment form 60 to 120 mm/rev the value of process time is decreased from 3.717 to 3.583 s, the further increment of C from 120 to 180 mm/rev, again process time is decreased from 3.583 to 3.167 s. This is due to the fact that machining time (process time) decreases with an increase in feed rate [21] for all brittle materials and no formation of chip clogging in-between tool and workpiece in CNC machine (Fig. 4.4c).

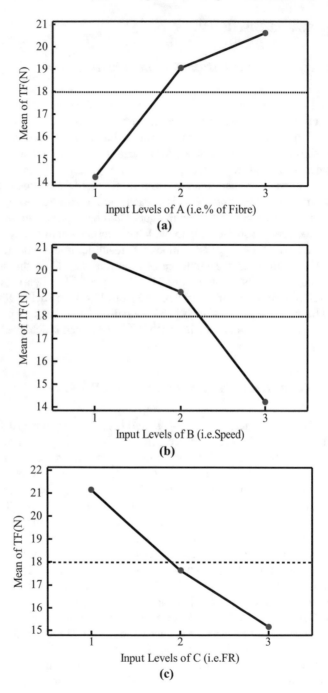

Fig. 4.3 a–c Main effect plots for thrust force versus input parameters levels

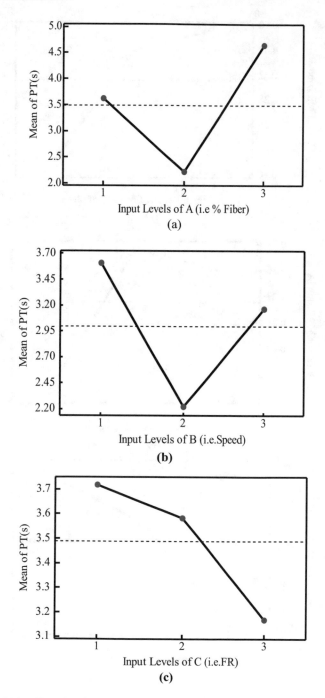

Fig. 4.4   a–c Main effect plots for process time versus input parameters levels

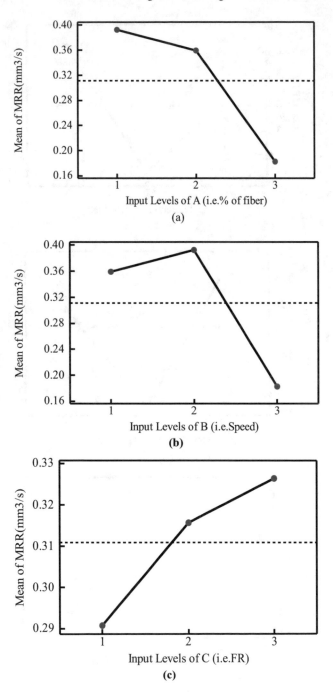

Fig. 4.5   a–c Main effect plots for MRR versus input parameters levels

4.3.3 Effects of Input Parameters on MRR

Similarly, the effect of input parameters A, B, and C on the MRR is performed, and the same is shown in Fig. 4.5. The plot shows that MRR decreased with an increase in A (% fiber) because higher % fiber concentration offers more resistance to material removal. With reference to the parameter B (i.e., speed), the MRR rate is increased from 0.359 to 0.392 mm³/s when the speed increased from 500 to 800 rpm (Fig. 4.5a). This is because proper cutting forces are acting on the work material, resulting in higher MRR. On the other hand, further increment of speed (B) form 800 to 1200 rpm the MRR shows decrement from 0.392 to 0.182 mm³/s because the gap between work material and tool increases and a further decrease in cutting forces [20, 21], resulting in lesser MRR (Fig. 4.5b). Moreover, the parameter C (i.e., feed rate) is positively influenced by MRR as shown in Fig. 4.6. It is observed from the main effect plot (Fig. 4.6) that, when the C (FR) increased from 60 to 80 mm/rev, the MRR is increased from 0.291 to 0.316 mm³/s further increment of parameter C (i.e., feed rate) form 120 to 180 mm/rev, and there is a slight increment of MRR from 0.316 to 0.326 mm³/s because tool penetration into the work material and cutting forces increases, resulting in higher material removal rate in the case of BFRP composite in CNC milling machining (Fig. 4.5c).

4.3.4 ANOVA

Additionally, a statistical analysis, i.e., ANOVA is performed to determine the significant parameters which affect the performance of CNC milling on BFRP composites. In this analysis, the relative importance of machining process parameters w.r.t. output responses to determine more accurate optimal combination machining parameters. Minitab 17 software is used for ANOVA and corresponding results are depicted in Tables 4.4, 4.5 and 4.6. In ANOVA, by using the stepwise elimination method, insignificant parameters are removed to adjust the fitted quadratic model [22–24].

The ANOVA result for thrust force as shown in Table 4.4 indicates that the parameter A, B (linear) and $B \times C$ (Interaction) found to be more significant for thrust force due to higher F-values and lesser P-value. The F-value of all these parameters obtained as 40.9, 35.51, 5.39 and corresponding P-values as 0.003, 0.004 and 0.021 < 0.05. Besides, sufficiency and suitability of the developed models are also tested by R^2 (coefficient of determination) and adjusted R^2. It is observed that higher R^2 value (95.62%) and adjusted R^2 (91.24%), shows the suitability and sufficiency of the model. Similarly, the result of ANOVA for process time of BFRP composites is tabulated in Table 4.5. The result indicates that parameters B (i.e., Speed) and C (i.e., feed rate) are found significant parameters on process time because of higher F-values and lesser P-values. The R^2 value obtained for process time is 81.04%, and the adjusted R^2 is 74.72% shows the suitability and sufficiency of the model [23]. At last, the ANOVA for MRR of BFRP composites is performed and a result is depicted

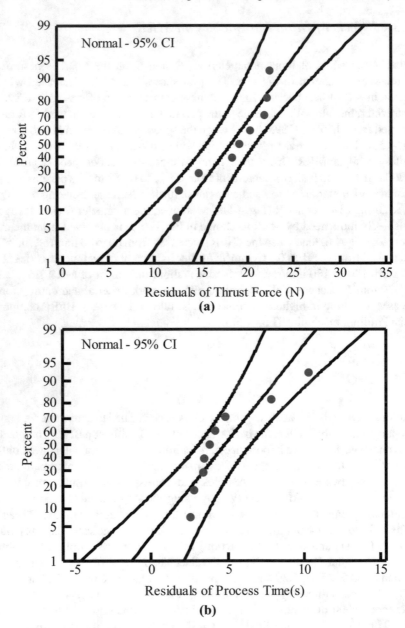

Fig. 4.6 Probability plot: **a** thrust force, **b** process time, and **c** MRR

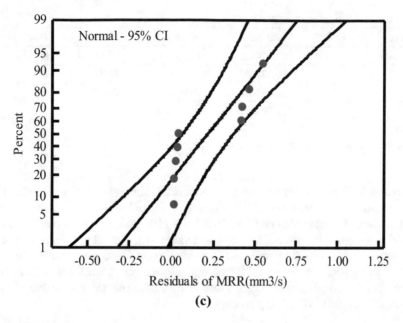

Fig. 4.6 (continued)

Table 4.4 ANOVA for thrust force of BFRP composite

Source	DF	Adj SS	Adj MS	F-value	P-value
Regression	4	129.581	32.395	21.84	0.006
A	1	60.664	60.664	40.9	0.003
B	1	52.663	52.663	35.51	0.004
C	1	25.745	25.745	17.36	0.064
B × C	1	7.995	7.995	5.39	0.021
Error	4	5.933	1.483		
Total	8	135.514			

$R^2 = 95.62\%$, R^2 (adj) $= 91.24\%$, R^2 (pred) $= 82.05\%$

Table 4.5 ANOVA of process time of BFRP composite

Source	DF	Adj SS	Adj MS	F-value	P-value
Regression	2	8.875	4.4374	12.82	0.007
A	1	6.687	6.6872	19.32	0.005
B	1	7.283	7.2835	21.05	0.004
Error	6	2.076	0.346		
Total	8	10.951			

$R^2 = 81.04\%$, R^2 (adj) $= 74.72\%$, R^2 (pred) $= 57.34\%$

Table 4.6 ANOVA of MRR of BFRP composite

Source	DF	Adj SS	Adj MS	F-value	P-value
Regression	2	0.076422	0.038211	38.13	0
A	1	0.072433	0.072433	72.28	0
B	1	0.074756	0.074756	74.6	0
Error	6	0.006013	0.001002		
Total	8	0.082435			

$R^2 = 92.71\%$, R^2 (adj) $= 90.27\%$, R^2 (pred) $= 83.59\%$

in Table 4.6. The results show that parameter A (% fiber) and B (Speed) found to be more significant parameters on MRR since the value of higher F-values and lesser P-values. Even, R^2 value obtained for MRR is 92.71%, and the adjusted R^2 is 90.27% is higher signifies the suitability and sufficiency of the model. Therefore, overall, parameters A (%fiber), B (speed), and C (feed rate) are found more significant on the CNC milling performance on BFRP composites. Hence, parameters A (%fiber), B (speed), and C (feed rate) are needed to take care of during the machining of BFRP composites in the CNC milling process.

4.3.5 Regression Analysis

To analyze the relationship between the input (A, B, and C) and output (thrust process, process time and MRR) parameters, a linear regression analysis has been carried out. In this analysis, the estimation of the value of one variable from the given value of another is done in Minitab-17 software. Here, three independent variables (A, B, and C) are taken for the analysis [23]. The mathematical relationship between A, B, and C and thrust force, process time, and MRR is an expression using Eq. (4.3).

$$Y = \alpha_0 + \alpha_1 X_1 + \alpha_2 X_2 + \alpha_3 X_3 \tag{4.3}$$

where Y is predictor (output) variable, i.e., MRR, TF, PT, etc. X_1, X_2 and X_3 are criterion (input) variables, i.e., A, B, C, and α_0, α_1, …, α_n are the regression coefficients.

In the analysis, experimental results (Table 4.3) are used and corresponding prediction equations are derived for MRR, thrust force and process time as a function of process parameters, i.e., % fiber, speed and feed rate as following:

$$TF = 46.09766667 - 7.7366 \times A + 0.11274 \times B - 0.05019444 \times C \tag{4.4}$$

$$MRR = 1.088 - 0.26733333 \times A + 0.0038666 \times B + 8.888 \times C \tag{4.5}$$

$$PT = 10.7633333 - 2.5686667 \times A + 0.003866667 \times B + 0.00458333 \times C$$
$$(4.6)$$

The impact on thrust force by various process parameters (A, B, and C) is shown in Eq. (4.4) and its reveals that B has a positive effect and A and C have a negative effect on thrust force and the process variable B is the most influencing process parameter. Similarly, the impact on process time by various process parameters is observed from Eq. (4.5), it is observed that variable B has a positive effect on process time while A and C have a negative effect on process time. Also, the impact on MRR by various process parameters is shown in Eq. (4.6) and it is observed that the variables B and C have a positive effect on MRR while A has a negative effect on MRR. From the regression analysis, parameters A and B have most significant on the CNC responses which justify the ANOVA results.

Further, the normality of residuals are plotted to determine the normality of the data points of thrust force, MRR, and process time using normality plots, and their graphical representation is shown in Fig. 4.6a–c. The graph shows that all points of the responses for thrust force, MRR, and process time are closer to the straight line and fallows normally distributed with 95% CI.

4.3.6 Modeling and Optimization

Modeling and optimization of the CNC milling process on the machining of BFRP composites are discussed. An integrated method, i.e., AHP with TOPSIS is employed [23] for modeling and optimization. In which, weights of the criteria, i.e., thrust force, process time and MRR is done using the AHP method while the optimal condition for the CNC milling process by TOPSIS method. Mainly, three output parameters, i.e., thrust force, MRR and process time are considered as criteria and 9 no's of experiments are considered as alternatives. In modeling first, the formulation of the decision matrix is designed using Eq. (4.7). It consists of performance values of criteria, i.e., thrust force, process time and MRR w.r.t. to alternatives, i.e., exp. number. Here, the experimental result (Table 4.3) is considered a decision matrix that is similar in the form of Eq. (4.7).

$$D_{ij} = \begin{bmatrix} & C_1 & C_2 & \ldots & C_n \\ A_1 & Y_{11} & Y_{12} & \ldots & Y_{1n} \\ A_2 & Y_{21} & Y_{22} & \ldots & Y_{2n} \\ \ldots & \ldots & \ldots & \ldots & \ldots \\ A_m & Y_{m1} & Y_{m2} & \ldots & Y_{mn} \end{bmatrix} \qquad (4.7)$$

where D_{ij} is the decision matrix contains response values of ith alternatives on jth criterion, i.e., Y_{11}, Y_{12}, ..., Y_{mn}; C_1, C_2, ..., C_n represents the no. of criteria or

Table 4.7 Normalized decision matrix

Exp. No.	TF in N	PT in s	MRR in mm³/s
1	0.487	0.486	0.019
2	0.208	0.208	0.043
3	0.235	0.235	0.452
4	0.157	0.157	0.496
5	0.258	0.258	0.028
6	0.298	0.298	0.037
7	0.212	0.212	0.447
8	0.171	0.172	0.588
9	0.64	0.64	0.017

no. of output parameters; $A_1, A_2, \ldots, A_m$ are a number of alternatives or experiments [22–24].

After that, normalization of the decision matrix is done using Eq. (4.8) to convert different measurements of units of criteria (i.e., thrust force, process time and MRR) into comparable sequences in the range of 0–1. The result of the normalized matrix is tabulated in Table 4.7.

$$N_{ij} = \frac{X_{ij}}{\left[\sum_{i=1}^{m} x_{ij}^2\right]^{1/2}} \quad \text{where } j = 1, 2, \ldots, n \qquad (4.8)$$

where N_{ij} represents normalized performance values of jth output parameters and ith experimental trial.

Then, the evaluation of criteria weights for each of the input parameters (i.e., thrust force, process time and MRR) is done using AHP. Here, the first pairwise comparison matrix between each criterion w.r.t each alternative i.e. A, B, and C is formulated in order to find comparative judgment between criteria and alternatives. The pairwise comparison matrix is developed based on the following expression [23].

$$x = \begin{bmatrix} P_{11} & P_{12} & \cdots & P_{1n} \\ P_{21} & x_{22} & \cdots & P_{2n} \\ \cdots & \cdots & \cdots & \cdots \\ P_{m1} & P_{m2} & \cdots & P_{mn} \end{bmatrix} \quad x_{ii} = 1, \; x_{ji} = \frac{1}{x_{ij}}, \; x_{ij} \neq 0 \qquad (4.9)$$

where P_{11}, P_{12}, Px_{1n} are the criterion and the major weights of two criteria are assigned using a fundamental scale of the AHP process as depicted in Table 4.8. The corresponding pairwise comparison matrix is tabulated in Table 4.9.

Based on the pairwise comparison matrix the principal eigenvector is determined and normalized. The relative priority weights (Table 4.8) are then computed from the comparison matrix by dividing the elements of each column by the sum of elements of the same column. Finally, the consistency of the pairwise matrix is calculated by

Table 4.8 Saaty AHP scale for pairwise comparison

Rating	Preferences
1	Equal importance
3	Equal importance
5	Strong importance
7	Very strongly importance
9	Extreme importance
2 4 6 8	Intermediate values between two adjacent judgments

Table 4.9 Pairwise comparison matrix and priority weight value of criteria

Criteria	TF in N	PT in s	MRR in mm^3/s	Eigen (λ) value	Priority (W$_i$)
TF in N	1	3	3	1.99	0.681
PT in s	0.334	1	3	0.79	0.117
MRR in mm^3/s	0.334	0.333	1	0.32	0.202
Total, Ti	5.143	14.533	22	8.08	1

finding the consistency ratio in order to judge the consistency of the values given by the decision-maker using Eqs. (4.10 and 4.11).

$$CI = \frac{(\lambda_{Max} - n)}{n - 1} \tag{4.10}$$

$$CR = \frac{CI}{RI} \tag{4.11}$$

where λ_{Max} is maximum eigenvalue, n is the number of the criterion being compared, random index (R.I.) is determined for a different order of pairwise comparison matrix, and its value is 0.9 for three criteria. The CR should be under 0.1 or less for a reliable result.

The results of the priority weight as depicted in Table 4.9 shows that the parameter or criteria thrust force found more significant followed by process time and MRR on the CNC milling performance because of higher priority weight. The CR value obtained for Table 4.9 is 0.09 < 0.1. Hence, the priority weights obtained from the pairwise comparison matrix are acceptable. In the fourth step, the weighted decision matrix is calculated using Eq. (4.12). This is formulated by multiplying the criteria weights of each CNC milling parameter (i.e., thrust force, process time and MRR) to each of the normalized experimental values as tabulated in Table 4.7. The normalized weighted matrix is depicted in Table 4.10 [23, 24].

$$U_{ij} = N_{ij} \times w_j \tag{4.12}$$

Table 4.10 Weighted normalized matrix

Ex. No.	TF in N	PT in s	MRR in mm³/s
1	0.1202	0.0474	0.0008
2	0.0514	0.0203	0.0017
3	0.058	0.0229	0.0178
4	0.0389	0.0153	0.0196
5	0.0637	0.0251	0.0011
6	0.0736	0.0291	0.0015
7	0.0524	0.0207	0.0177
8	0.0423	0.0167	0.0232
9	0.158	0.0624	0.0007

where U_{ij} is a weighted normalized decision matrix, w_j is the criteria weights of each of the input parameters.

The closeness coefficient (CC) values for each of the alternatives, i.e., exp. no is calculated using Eq. (4.17) followed by positive separation (U^+) and negative separation (U^+) and positive separation ideal solution (S^+), negative separation ideal solution (S^-) for each of the criteria using Eqs. (4.13–4.16). These are calculated to determine the relative distance between criteria and alternatives with considering the higher-the-better and lower-the-better conditions of the criteria [23, 24].

Maximum value

$$U^+ = \left\{ \left[\overset{\text{max}}{\underset{i}{\sum}} U_{ij} | j \in J \right], \left[\overset{\text{min}}{\underset{i}{\sum}} U_{ij} | J \in j | i = 1, 2, 3, \ldots, m \right] \right\} \quad (4.13)$$

$$U^+ = \left(U_1^+, U_2^+, \ldots, U_n^+ \right) \text{ Maximum value}$$

Minimum value

$$U^- = \left\{ \left[\overset{\text{min}}{\underset{i}{\sum}} U_{ij} | j \in J \right], \left[\overset{\text{max}}{\underset{i}{\sum}} U_{ij} | J \in j | i = 1, 2, 3, \ldots, m \right] \right\} \quad (4.14)$$

$$U^- = \left(U_1^-, U_2^-, \ldots, U_n^- \right) \text{ Minimum value}$$

Positive ideal and negative ideal solution

$$S_i^+ = \sqrt{\sum_{j=1}^{n} \left(U_{ij} - U_j^+ \right)^2}, \ i = 1, 2, 3, \ldots, m \quad (4.15)$$

Table 4.11 Ranking of CC values for CNC milling process

Exp. No.	CC values	Rank
1	0.448	8
2	0.451	7
3	0.459	5
4	0.455	6
5	0.446	9
6	0.474	3
7	**0.523**	**1**
8	0.472	4
9	0.523	2

Bold term indicates the sequence of ranking

$$S_i^- = \sqrt{\sum_{j=1}^{n} \left(U_{ij} - U_j^- \right)^2}, \quad i = 1, 2, 3, \ldots, m \qquad (4.16)$$

$$CC = \frac{S_i^-}{S_i^- + S_i^+}, \quad i = 1, 2, 3, 4, \ldots, m \qquad (4.17)$$

where S_i^+ = Euclidean distances positive ideal solution, S_i^- = Euclidean distances negative ideal solution, CC is the closeness coefficient values for each of the alternatives.

At last, rank the optimal alternative, i.e., optimal experimental setting or exp. no based on the closeness coefficient values (CC) is done. The choice with greater closeness coefficient values gives optimal operating conditions to get higher performance for the CNC milling system compared to the other experiment number. The outcomes of the ranking are tabulated in Table 4.11.

The result shows that Exp. No. 7 gave higher closeness coefficient values (0.543) and found to be the optimal setting for the CNC milling process among the other experimental runs. The optimal setting obtained based on Exp. No. 7 is fiber % (A as 10%, level 1), speed (B as 500 rpm, level 1), and feed rate (C as 180 mm/rev, level 3). The optimal parameter obtained using AHP-TOPSIS method offers the most desirable output parameter such as higher MRR, lesser thrust force, and process time which directly or indirectly ensures product quality improvement, reduction in manufacturing cost, and machining efficiency enhancement of the CNC milling process on machining of BFRP composites. Hence, work suggested that ideal setting, i.e., fiber % (A as 10%, level 1), speed (B as 500 rpm, level 1), and feed rate (C as 180 mm/rev, level 3) can be used as a standard for CNC milling for effective machining of BFRP composites. Furthermore, the proposed method, i.e., AHP-TOPSIS can be utilized for finding the best possible combination of input parameters for CNC milling process and other processes as well.

Table 4.12 Confirmatory tests for CNC milling process of BFRP composites

Input conditions	Output parameters	Experimental results	Confirmatory tests results
Exp. No. 7 fiber % (*A* as 10%), speed (*B* as 500 rpm) and fed rate (*C* as 180 mm/rev)	TF (N)	14.302	13.765
	MRR (mm³/s)	0.373	0.302
	PT (s)	3.650	3.342

4.3.7   Confirmatory Analysis

Additionally, the justification of results obtained from the AHP-TOPSIS method is done through the confirmatory analysis. For this, optimal setting [i.e., fiber % (*A* as 10%), speed (*B* as 500 rpm), and feed rate (*C* as 180 mm/rev)] obtained using the proposed method is considered and same being re-tested in CNC milling process. It has been observed from the results (Table 4.12) that confirmatory test results found to be comparable and acceptable with the experimental results for the selected optimal setting [22–24].

4.4   Summary

Today, with the need of more flexible material and environmental impact, present industries started to use natural fiber-based polymer composites due to their greater specific strength, eco-friendly, biodegradable, and low cost. In fact, use of banana fiber in manufacturing of polymer composites is also increased recently for various applications. But these composites undergo various secondary operations like drilling, milling, facing, etc. before actual use in assembly process. Hence, the present chapter discussed the manufacturing and processing/machining of BFRP composites. Here, first fabrication of three different % of fibers, i.e., 10, 15 and 20% composites are prepared by hand layup process. Then machining on these composites is performed using CNC milling. Work considered % fiber, speed and feed rate as input variables while MRR, thrust force and process time as output variables. Result shows that hand layup technique is more effective in production of BFRP composites compared to others. Parameter % fiber and speed found more significance in the performance of CNC milling process. Optimal setting obtained by AHP-TOPSIS is at fiber % (*A* as 10% with level 1), speed (*B* as 500 rpm with level 2), and feed rate (*C* as 180 mm/rev with level 2) which provides optimal responses as thrust force (14.302 N), MRR (0.373 mm³/s), process time (3.650 s), respectively. Additionally, ANOVA and empirical models for process time, thrust force, and MRR of BFRP composites found statistically significant and fit the data well enough with the experiment results with the 95% confidence level. At last, confirmatory tests are done to justify the results and found very much comparable and acceptable with the experimental results.

Based on the above observations, the banana fiber-reinforced polymer composites give better execution in the CNC milling process. The parametric settings got from the AHP-TOPSIS technique can be utilized as the ideal setting for the CNC milling during the machining of banana fiber-reinforced polymer composites. Moreover, the created empirical model for CNC milling responses can be implemented in an organized structural model for prediction of process time, thrust force, and MRR for all the cases.

References

1. B. John, A.K. Amir, E. Alaa, B. Ahmed, Production, application and health effects of banana pulp and peel flour in the food industry. J. Food Sci. Technol. **56**(2), 548–559 (2019)
2. G. Amaresh, G.K. Praveennath, M. Nagamadhu, S.B. Kivade, K.V.M.S. Murthy, Study on mechanical properties of alkali treated plain woven banana fabric reinforced biodegradable composites. Compo. Commun. **13**, 47–51 (2019)
3. K. Senthilkumar, I. Siva, J.W.J.M. Vikneshwararaj, P. Karthick, P. Devakumar, Influence of orientation on tensile and flexural properties of sisal fiber polyester composite. J. Chem. Pharm. Sci. **974**, 2115 (2015)
4. K. Susheel, B.S. Kaith, K. Inderjeet, Pretreatments of natural fibers and their application as reinforcing material in polymer composites-a review. Poly. Eng. Sci. **49**, 1253–1272 (2009)
5. A.V.R. Prasad, K.M. Rao, Mechanical properties of natural fiber reinforced polyster composites: Jowar, Sisal and Bamboo. Mater. Des. **32**(8–9), 4658–4663 (2011)
6. R.K. Misra, S. Kumar, static and dynamic mechanical analysis of chemically modified randomly distributed short banana fiber reinforced high density polyethylene/poly (ε-caprolactone) composites. Soft Mater. **4**(1), 1–13 (2007)
7. J.S.M. Osorio, R.R. Baracaldo, J.J.O. Florez, The influence of alkali treatment on banana fibre's mechanical properties. Ingeniería Invest. **32**(1), 83–87 (2012)
8. A. Alavudeen, N. Rajani, S. Karthikeyan, M. Thiruchitrabalam, N. Venkateshwaren, Mechanical properties of banana-kenaf fiber reinforced hybrid polyester composites: effect on woven fabric and random orientation. Mater. Des. **66**, 246–256 (2015)
9. M.H. Md Radzi, K. Abdan, A.Z. Zainal, M.A. Md Deros, Zin, Thermal and flammability characteristics of blended jatropha bio-epoxy as matrix in carbon fiber-reinforced polymer. J. Comp. Sci. **3**(1), 3390–33100 (2019)
10. T.P. Mohan, K. Kanny, Compressive characteristics of unmodified and nanoclay treated banana fiber reinforced epoxy composite cylinders. Compo. Part B: Eng. **169**, 118–125 (2019)
11. S. Jayabal, U. Natarajan, Optimization of thrust force, torque, and tool wear in drilling of coir fiber-reinforced composites using Nelder-Mead and genetic algorithm methods. Int. J. Adv. Manuf. Technol. **51**, 371–372 (2010)
12. S. Jayabal, U. Natarajan, Drilling analysis of coir-fibre-reinforced polyester composites. **34**(7), 1563–1567 (2011b)
13. A. Athijayamani, U. Natarajan, M. Thiruchitrambalam, Prediction and comparison of thrust force and torque in drilling of natural fibre hybrid composite using regression and artificial neural network modelling. Int. J. Mach. Mach. Mater. **8**, 131–145 (2010)
14. D. Chandramohan, K. Marimuthu, Drilling of natural fiber particle reinforced polymer composite material. Int. J. Adv. Eng. Res. Stud. **1**(1), 134–145 (2011)
15. K. Patel, Investigation on drilling of banana fibre reinforced composites, in *2nd International Conference on Civil, Materials and Environmental Sciences*, pp. 201–205 (2015)
16. F.M. AL-Oqla, S.M. Sapuan, Natural fiber reinforced polymer composites in industrial applications: feasibility of date palm fibers for sustainable automotive industry. J. Clean. Prod. **66**, 347–354 (2014)

17. G.D. Babu, Y. Kasu, Determination of delamination of milled natural fiber rainforced composits. Int. J. Eng. Res. Technol. **1**(8), 1–6 (2012)
18. N. Venkateshwaran, A. Elaya Perumal, D. Arunsundaranayagam, Fiber surface treatment and its effect on mechanical and visco-elastic behaviour of banana/epoxy composite. Mater. Des. **47**, 151–159 (2013)
19. H.R. Maleki, M. Hamedi, M. Kubouchi, Y. Arao, Experimental study on drilling of jute fiber reinforced polymer composites. J. Compo. Mater. **0**(0), 1–13 (2018)
20. J. Kechagias, G. Petropoulos, Application of Taguchi design for quality characterization of abrasive water jet machining of TRIP sheet steels. Int. J. Adv. Manuf. Technol. **62**, 635–643 (2012)
21. M. Mudhukrishnan, P. Hariharan, K. Palanikumar, Measurement and analysis of thrust force and delamination in drilling glass fiber reinforced polypropylene composites using different drills. Measurement **149**, 106973 (2020)
22. M.J. Davidson, K. Balasubramanian, G. Tagore, Surface roughness prediction of flow-formed AA6061 alloy by design of experiments. J. Mater. Process. Technol. **202**, 41–46 (2008)
23. A.V.S.R. Prasad, K. Ramji, M. Kolli, G.V. Krishna, Multi response optimization of machining process parameters for wire electric discharge machining of lead induce Ti–6Al–4 V using AHP-TOPSIS method. J. Adv. Manuf. Syst. **18**(2), 213–236 (2019)
24. S. Bhowmik, Jagadish, K. Gupta, Modeling and optimization of advanced manufacturing processes, in *Manufacturing and Surface Engineering*, 1st edn (Springer International Publishing, 2019), p. 1–74. https://doi.org/10.1007/978-3-030-00036-3

Chapter 5
Manufacturing and Processing of Jute Filler-Based Polymer Composite

Abstract This chapter addresses the manufacturing and processing or machining aspects of Jute fiber-based polymer (JFRP) composites using AWJM machining is a commonly known as green machining process. First, composite with 5% wt% of fiber is fabricated using hand layup method followed by compression molding. Second, experiments are executed as per the Taguchi-based L_9 orthogonal array to assess the influence processing attributes [stand of distance, working pressure; speed of the nozzle] on MRR and surface quality. Third, parametric analysis, ANOVA and multi-regression analysis are also done to show the statistical significance of the processing data and also optimized within the tested range through MOOSRA method. Investigation result indicates that AWJM is capable in machining/processing of JFRP composites with better quality. Ideal setting obtained is stand of distance as 3 mm, working pressure as 100 MPa and nozzle speed as 300 mm/min and variable speed of the nozzle and stand of distance are found more significance on MRR and surface quality in all the cases. Last, optimum parameters levels are verified via confirmatory test and surface morphology and found acceptable values to the executed data and smooth & uniform machined surface.

Keywords JFRP composite · AWJM · Parametric analysis · ANOVA · Regression analysis · Optimization · MOOSRA · Surface integrity

5.1 Introduction

Jute is the oldest fiber crop (*Corchorus* spp.) that belongs to *Tileaceae* and *Genus corchorus* family. Jute is also named 'golden fiber'. Kalpana and Devegowda [1] explored on production and productivity of jute in India. They found that an estimated area of 7.9 lakh hectares with an average production rate of 102.85 lakh bales (1 bale = 180 kg) are produced. Mostly, in India, around 4 million farm families are engaged in jute farming. Almost all states in India jute is cultivated, as per the national agricultural department (NAD) reports that the maximum jute is cultivated in West Bengal around 74 and 81.6% of national acreage and production. White jute (*corchorus capsularis*) and tossa jute (*corchorus olitorius*) are the two most

Jagadish and S. Bhowmik, *Manufacturing and Processing of Natural Filler Based Polymer Composites*, SpringerBriefs in Applied Sciences and Technology,
https://doi.org/10.1007/978-3-030-65362-0_5

commercial varieties of species of corchorus family. One of the most advantages in jute plant is every part of the plant is useful, viz. fiber (bast), stake, seed, and lends, out of which bast is a most beneficial part of the jute plant and the threads(fibers) are extracted from the stem of the jute plant can be utilized in many applications like clothing, construction, paper, health, foods, body care, textiles, medical, packing, cosmetics, and bio-fuels so that it is really eco-friendly fiber. In present days jute fibers are widely used in automobile industries because of their high impact, tensile, low moisture absorption, and flexural properties (i.e., high mechanical properties). The industrial term for jute fiber is also called as raw jute. The jute fiber is off-white to brown in color, and fiber is grown approximately between 1 m (3 ft) and 4 m (13 ft) long, based on the method of extraction of the fibrils from the stem. The chemical composition of jute fibers consists of hemicelluloses (20–24%), cellulose (58–63%), and lignin (12–15%). Because of the high content of cellulose, the degree of crystalline, jute fibrils is selected as reinforcement material in the present work.

Instead, jute fibers are eco-friendly, less in density, and economically cheap and have larger industrial applications compared to the other fibers [hemp, coir, flax, coconut, sisal, and banana, bamboo, bagasses, rice husk, etc.]. Meanwhile, some extent of work has been done on the jute fiber-based composites with their physical and mechanical properties. Lenord and Ansell [2] presented work on chemical treatment of natural fibers (jute, sisal, hemp, and kapok) using NaOH solution. The result shows that the mechanical properties are improved, increases the crystallinity index and thermal stability. Singh et al. [3] considered the fabrication and mechanical properties of JFRP composites. The result shows that the chemical treatments, fiber orientation, and fiber composition are major elements to improve the mechanical properties of JFRP composite. Prasad et al. [4] investigated the Jute and banana fiber-reinforced polymer hybrid composite properties using FEM. The result shows that mechanical properties were enhanced and replaced synthetic resins. Kaushik et al. [5] investigated on mechanical properties of jute fiber epoxy/polyester composites. Gowda et al. [6] assessed the mechanical characteristics of reinforced jute fabric composites of polyester and found that they are more effective than Composites focused on wood. The tensile and flexural properties were examined by Luo and Netravali, in which property of pineapple fiber compared with pure resin with composites [7]. Karmaker and Schneider [8] have fabricated jute and kenaf fiber composite, containing polypropylene resin as matrix material. They found that jute fiber exhibits better mechanical properties than kenaf fibers. Tensile, bending and impact tests were conducted by Chandrasekhar et al. to investigate the orientation effect on bamboo laminates. Hojo et al. [9] tested tensile property jute, kenaf and bamboo mat composite. Tensile modulus and ultimate strength of bamboo and jute are similar, while kenaf show high tensile modulus and ultimate strength than jute and bamboo. The results clearly observed that the epoxy-based jute polymer composite had good mechanical properties as compared to polyester-based jute composite. However, polyester-based jute composite has a high wear rate as compared to epoxy-based jute polymer composite.

On the other hand, a JFRP composite undergoes various secondary operations before going to the actual use. Secondary operations include drilling, trimming, cutting, and milling to get the final product. Thus, machining especially drilling, trimming, cutting, and milling with greater productivity at least machining costs, unlike shape with high superior finish and high machining effectiveness, has become imperative to study the machinability of JFRP composite. In fact, machining of JFRP composite is found to be quite challenge and costly attempt using conventional machining due to their mechanical anisotropy and heterogeneous properties. Subsequently, the conventional machining process produces an extreme heat generation at the cutting zone and problems in heat dissipation due to the comparatively least heat conductivity of the materials. As a result, it decreases the tool life, increases the surface roughness, and decreases the dimensional sensitivity of work materials [10–21]. Apart from the heat generation, the conventional machining, i.e., drilling process generates various environmental harmful components or poisonous constituents in the form of waste which outcomes in severe health and environmental issues. To resolve these, non-conventional machining such as AWJM, EDM, USM, and WEDM process is employed. Since all the non-conventional machining processes are not suitable for machining JFRP composites, the care-full selection of non-conventional machining is therefore essential. Hence, in this chapter, the considered AWJM method is used for the machining of JFRP composites, because this process is suitable for machining all kinds of materials including ductile, brittle, etc. It offers many advantages such as greater flexibility in machining, reduced waste materials, less heat affected zone, minimal force during cutting, less contamination, no lubricant required, and no cooling required. Hence, this process is sometimes called a green manufacturing process. However, no work in the literature on the feasibility of using the AWJM process for machining of JFRP composites is avilable.

5.2 Materials Manufacturing and Machining Details

5.2.1 Manufacturing of JFRP Composites Using Compression Molding Technique

In this work, jute fibers are extracted from jute stems (*corchorus olitorius*). These stems are dried for 2–3 days within the cropland and drowned in water for duration of one week to allow for bacterial fermentation to take place. After fermentation, wet stems are removed from the water and dried into the sunlight for a period of 7 days, then after peeled fibers from the stem and washed with clean distilled water. Further, moisture present in the jute fiber is removed by keeping in woven around 12 h. Later, jute fiber cut into small pieces and converted into powder form with particle size is 2 μm and reinforced material weigh percentage is 5, 7.5 and 10% in matrix material, i.e., Epoxy resin (Araldite LY 556) with the density of 1.26 gm/cc and equivalent hardener (HY 951) in the ratio of 10:8 (by weight %). After that, mixer of fiber and

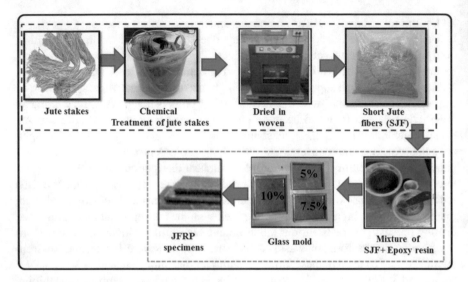

Fig. 5.1 Fabrication steps of JFRP composites

epoxy properly stirred and slowly poured in glass mold (Fig. 5.1). To remove the moisture present in the mixture is cured for 24–48 h at chamber temperature. Lastly, three different specimens (Fig. 5.1) of 5, 7.5 and 10% of size 80 mm × 40 mm × 5 mm are considered for machining [18]. The detailed fabrication steps of JFRP composite is illustrated in Fig. 5.1.

5.2.2 Machining Details

The drilling experiments are conducted on a 5% JFRP sample by using computer numerical controlled AWJ Cutting Machine, as exposed in Fig. 5.2a, b. Based on the experimental setup, a total of 9 holes are drilled on a 5% JFRP composite sample, using a working pressure of 100–150 MPa, stand of distance of 1–3 mm, nozzle speed of 100–300 mm/min and jet of angle fixed at 90°. The orifice is used to control the flow with a flow rate of 5 g/s and having a diameter of 0.25 mm [19]. Distill water mixed with garnet [80 mesh size] is used as working slurry and voltage 300 V and current 20 A is kept as constant in machining.

The stages of independent parameters mainly working pressure (WP), nozzle speed (NS), and standoff distance (SOD) are nominated according to previous works and existing AWJM setup. The influences on dependent parameters are such as surface roughness (SR) and material removal rate (MRR) of JFRP composite by the AWJM process. Parameter details and experimental design as per the Taguchi (L9) orthogonal array are depicted in Tables 5.1 and 5.2, respectively.

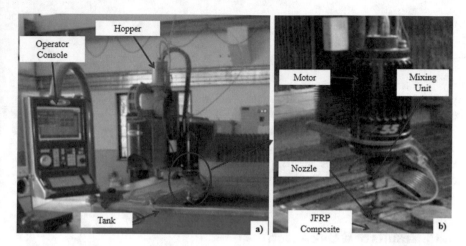

Fig. 5.2 a AWJ machine, **b** AWJM head setup

Table 5.1 AWJM process parameters [33]

Input parameters	Stage 1	Stage 2	Stage 3
WP in MPa	100	125	150
SOD in mm	1	2	3
NS mm/min	100	200	300

Table 5.2 Trial run results of AWJM process

Exp. No.	% Composition	Input parameters			Output parameters	
		SOD (mm)	WP (MPa)	NS (mm/min)	MRR (g/min)	SR (μm)
1	5%	1	100	100	3.384	0.118
2		1	125	200	12.029	0.167
3		1	150	300	26.619	0.182
4		2	100	200	22.573	0.198
5		2	125	300	42.826	0.202
6		2	150	100	16.559	0.225
7		3	100	300	57.274	0.231
8		3	125	100	21.831	0.253
9		3	150	200	48.918	0.275

During the experiment, three process parameters (shown in Table 5.1) are varied according to the experimental design, and a square hole of 20 mm is cut. Every trial run is executed 3 times and their average of MRR and surface roughness is taken for the analysis. These trials are conducted to know the influence of individual factors of experimental design as shown in Table 5.2. The output parameter MRR is estimated using the following equation:

$$MRR = \frac{I_m - F_m}{\rho \times PT}$$ (5.1)

where I_m and F_m are the initial and final masses of JFRP composite; PT is cutting/process time, ρ is the density of the JFRP composite.

After machining, machined surface is measured by Surface Profilometer (Make: Tokyo Seimitsu Co. Ltd. Model: Handysurf E-35B). Surface roughness is calculated at three different locations. Similar procedure is followed for four side faces and their average values with lower-the-better conditions are taken for the analysis (Table 5.2).

5.3 Results and Discussion

5.3.1 Influence of Process Parameters on MRR

During the machining operation, the response parameter MRR is studied at three different process factors such as SOD, working pressure and nozzle speed as shown in Fig. 5.3. It is witnessed that all the process factors significantly effect on the response parameter MRR, i.e., increase in SOD, working pressure, and nozzle speed, resulting in a major increase in MRR during machining of JFRP composite [22]. It is also observed from Fig. 5.3 is that the response parameter MRR decreases with increasing the fiber composition. This is because; the more fiber content in the composite offers high resistance to the removal of material. It is clearly observed

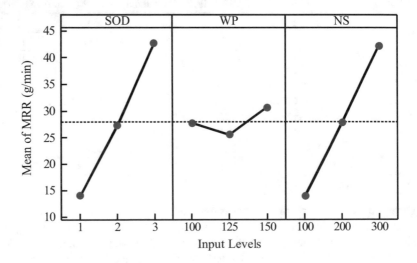

Fig. 5.3 Main effect plot for MRR versus input parameters levels

from Fig. 5.3 that the independent parameter SOD is varied from 1 to 3 mm, and then the response parameter MRR is significantly increased in the case of 5% fiber composition. This is because the process parameter SOD rises, and the radius of jet occupancy on the work material is also increased, resulting in higher MRR. Also, a similar pattern is observed from Fig. 5.3 that the process parameter nozzle speed varies from 100 to 300 mm/min. It is also identified from Fig. 5.3 that, in the case of 5% fiber composition graph shows, the input parameter working pressure rises from 100 to 125 MPa, the dependent parameter is marginally decremented from 27.8 to 25.5 g/min, and further, the response parameter MRR is considerably increased from 25.5 to 30.8 g/min with increasing the parameter working pressure from 125 to 150 MPa. This is because the process parameter working pressure and nozzle speed increase and also increases the kinetic energy of the abrasive particles inside the nozzle, and then the high-velocity jet of abrasive particles contact to the surface of the work material results in higher MRR [23]. Hence, it comes to know that the process parameter SOD, working pressure, and nozzle speed are detected to be the most influencing factors on MRR for AWJM process during processing of JFRP composites. Based on this analysis, the optimal combination of process factors JFRP composites is at stand of distance (3 mm, stage 3), working pressure (150 MPa, stage 3), nozzle speed (300 mm/min, stage 3).

5.3.2 Influence of Process Parameters on Surface Roughness

While the machining of JFRP composite by using the AWJM process, the impact of self-governing factors such as SOD, working pressure, and nozzle speed on surface roughness of JFRP composite is analyzed [24] as shown in Fig. 5.4. It is clearly identified that the input parameters SOD and working pressure have a positive effect on response parameter surface roughness and independent parameter nozzle speed shows the minimal effect on surface roughness. And it is also observed from Fig. 5.4 that, for 5% fiber composition graph shows, the process parameter SOD varies from 1 to 3 mm, and the response parameter surface roughness is significantly raised from 0.1556 to 0.2083 μm. This is due to, at maximum SOD, the kinetic energy which also increases, and it helps to remove the material quite easily with higher surface roughness [24, 25]. Also, an increase of parameter working pressure results in positive increases of response parameter surface roughness in the case of 5% fiber composition can be seen in Fig. 5.4. As the working pressure varies from 100 to 150 MPa, the response parameter surface roughness is significantly raised from 0.1556 to 0.2265 μm. This is because if the parameter working pressure increases gradually, the impact of abrasive particles on work material also increases, resulting in and higher surface roughness which is obtained. And also, it is observed from Fig. 5.4 that, for 5% fiber composition of JFRP graph shows, the independent parameter nozzle speed increases from 100 to 200 mm/min, and the dependent parameter surface roughness is gradually raised from 0.198 to 0.214 μm, while the output parameter surface roughness value drastically decreases from 0.214 to 0.204 μm with increasing

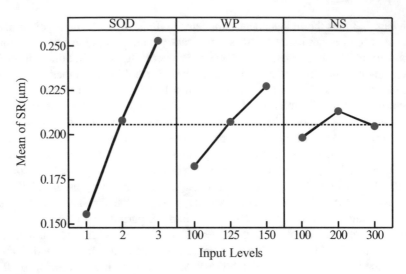

Fig. 5.4 Main effect plot for surface roughness versus input parameters levels

the nozzle speed from 200 to 300 mm/min. This is due to the reason that at lower nozzle speed (100 mm/min), lesser material is removed, and hence, better value of surface roughness is obtained. Based on this analysis, the optimal combination of process factors JFRP composites is at stand of distance (1 mm, stage 1), working pressure (100 MPa, stage 1), nozzle speed (100 mm/min, stage 1).

5.3.3   ANOVA

Analysis of variance (ANOVA) is discussed to study the effect of each parameter of AWJM process on jute-based reinforced polymer composites while machining with the AWJM process. Minitab17 version software is used for ANOVA and the outcomes are tabulated in Tables 5.3 and 5.4. In ANOVA, first identify significant parameters; remaining insignificant parameters removed from the table and to adjust the fitted quadratic model [26]. If the parameters value lesser than the P-value (probability value) then the parameters are said to be significant. Similarly, the value of F is more, and it seems that the process parameters performance characteristics also change [27]. Table 5.3 shows the ANOVA results for MRR. It has been observed that the P-value of interaction SOD with nozzle speed and working pressure with nozzle speed is smaller, i.e., F-value of interaction SOD with nozzle speed is larger at 241.18 and P-value is exactly 0.004, and also, the interaction working pressure with nozzle speed of F-value is higher at 27.93 and P-value is 0.034. Moreover, the F-value of nozzle speed is larger at 16.68 and P-value is 0.05. This represents P-values are lesser than 0.05. This is because the interaction parameters SOD with nozzle speed, working pressure with nozzle speed, and nozzle speed are the most

Table 5.3   ANOVA for MRR of JFRP composite

Source	DF	F-value	P-value
Regression	6	1548.94	0.001
SOD	1	0.23	0.68
WP	1	0.31	0.633
NS	1	16.68	0.05
SOD × NS	1	241.18	0.004
WP × NS	1	27.93	0.034
SOD × WP	1	0.37	0.607
Error	2		
Total	8		

R^2: 99.98%, R^2 (adj): 99.91%, R^2 (Pred): 99.67%

Table 5.4   ANOVA for surface roughness of JFRP composite

Source	DF	F-value	P-value
Regression	3	54.78	0
SOD	1	8.01	0.037
WP	1	8.58	0.033
SOD × WP	1	0.95	0.375
Error	5		
Total	8		

R^2: 97.05%, R^2 (adj): 95.28%, R^2 (Pred): 85.41%

significant parameters on response parameter MRR. The value of R^2 obtained for MRR is 99.98% and the adjusted R^2 is 99.91%, this value represents that the present model data is in line with the [28].

Similarly, ANOVA result for surface roughness is carried out and the outcomes of the same are represented in Table 5.4, for 5% fiber composition of JFRP composite shows that the process factors SOD and working pressure are utmost affecting factors on response parameter surface roughness with higher F-values and smaller P-values. The value of R^2 obtained for surface roughness is 97.05% and the adjusted R^2 is 95.28% which is an indication that the present model data fit well enough.

5.3.4   Empirical Model

In this section, empirical model development has been discussed. Using multiple regression analysis in Minitab software, individual empirical models have been generated and plotted. The individual empirical model consists of a set of Eqs. (5.2–5.3) comprising process parameters such as SOD, working pressure, and nozzle speed and response factors, viz. MRR, and surface roughness [29].

$$MRR = 0.82 - 0.99 \, SOD - 0.0192 \, WP - 0.1371 \, N + 0.07568 \, SOD \times NS$$
$$+ 0.001030 \, WP \times NS + 0.0118 \, SOD \times WP \tag{5.2}$$

$$SR = -0.0866 + 0.0368 \, SOD + 0.001057 \, WP + 0.001037 \, NS$$
$$- 0.000156 \, SOD \times NS - 0.000005 \, WP \times NS$$
$$+ 0.000291 \, SOD \times WP \tag{5.3}$$

The empirical models (Eqs. 5.2–5.3) will be used for forecast of the response factors, i.e., MRR and surface roughness for the different input conditions of the AWJM process in machining of JFRP composites. Further, the normality of residuals are plotted to determine the normality of the data points of MRR, and surface roughness using normality plots and their graphical representation are shown in Fig. 5.5a, b. The graph shows that all points of the responses for MRR and surface roughness are in line with the straight line. Therefore, it is seen that the experimental data of MRR and surface roughness are normally distributed well within the 95% CI.

5.3.5 Modeling and Optimization

In this section, multi-response optimization of AWJM process on machining of JFRP composites is discussed using multi-objective optimization based on the simple ratio analysis (MOOSRA) [30–32]. The parameters MRR and surface roughness are considered as output parameters while SOD, working pressure, and nozzle speed are process parameters. In this method first, design a decision matrix is carried out using Eq. (5.4). Taguchi (L_9) is used as base for creation of decision matrix in which experiment trials (from 1 to 9) are taken as alternatives (i.e., A_1, A_2, …., A_n) and output parameters as criteria (C_1, C_2, and C_3). After formulation, the decision matrix is tabulated in Table 5.5.

$$D_{ij} = \begin{bmatrix} & C_1 & C_2 & \dots & C_n \\ A_1 & Y_{11} & Y_{12} & \dots & Y_{1n} \\ A_2 & Y_{21} & Y_{22} & \dots & Y_{2n} \\ \dots & \dots & \dots & \dots & \dots \\ A_m & Y_{m1} & Y_{m2} & \dots & Y_{mn} \end{bmatrix} \tag{5.4}$$

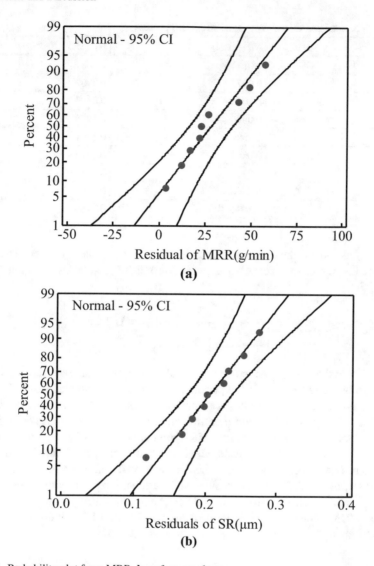

Fig. 5.5 Probability plot for **a** MRR, **b** surface roughness

where D_{ij} is the decision matrix contains response values of ith alternatives on jth criterion, i.e., Y_{11}, Y_{12}, ..., Y_{mn}; C_1, C_2 ..., C_n represents the number of criteria or no. of response parameters; A_1, A_2 ... *is* a no. of alternatives or experiments.

Normalization of the decision matrix carried out by using Eq. (5.5) to convert the different measurement data of performance values into a single response parameter and compare sequence data. This step is essential because processing of JFRP composites in AWJM problem contains different units of output parameters like MRR and SR [31]. In order to evaluate the overall assessment for AWJM process,

Table 5.5 Normalized values (N_{ij}) values of AWJM process

Exp. No.	N_{ij} (MRR)	N_{ij} (SR)
1	0.0134	0.0637
2	0.0477	0.0902
3	0.1056	0.0983
4	0.0896	0.1070
5	0.1699	0.1091
6	0.0657	0.1216
7	0.2273	0.1248
8	0.0866	0.1367
9	0.1941	0.1486

one needs to convert these output parameters into comparable sequence. The result of normalization matrix for AWJM process on JFRP composites is tabulated in Table 5.5.

$$N_{ij} = \frac{D_{ij}}{\left[\sum_{i=1}^{m} D_{ij}^2\right]^{1/2}} \text{ where } j = 1, 2, \ldots, n \tag{5.5}$$

Thereafter, the overall assessment values of the output parameters (i.e., MRR, and surface roughness) for each of the parameter settings are evaluated using Eq. (5.6). The final assessment values of response parameters out of which each output parameter values alter the multi-response optimization into a single response optimization problem. The output parameters like MRR are considered as beneficial parameters while surface roughness is considered as non-beneficial parameter. Based on the assessment (y_j) values, ranking is done for each of the alternatives (exp. no) and optimal setting for AWJM process is established. The exp. no with higher assessment (y_j) values yields the first rank and corresponding setting values are the finest for the AWJM process at some point of the machining of JFRP composites. The results of the optimized parameters are shown in Table 5.6.

$$y_j = \sum_{i=1}^{g} N_{ij} \div \sum_{i=g+1}^{n} N_{ij} \tag{5.6}$$

where N_{ij} is the normalized decision matrix for beneficial criteria and non-beneficial criteria, g is the no. of criteria that are maximization condition, $(n - g)$ is the no. of criteria that are minimization condition, and y_j is the assessment values for ith criteria with respect to all jth exp runs.

From Table 5.6, Exp. No. 7 shows maximum assessment (y_j) values for JFRP composite and corresponding ideal conditions obtained as stand of distance (3 mm, stage 3), working pressure (100 MPa, stage 1) and nozzle speed(300 mm/min, stage

Table 5.6 Assessment values (y_j) values of AWJM process

Exp. No.	Input variables			y_j values	Rank
	SOD (mm)	WP (MPa)	NS (mm/min)		
1	1	100	100	0.221	9
2	1	125	200	0.387	8
3	1	150	300	0.560	5
4	2	100	200	0.544	6
5	2	125	300	0.756	3
6	2	150	100	0.525	7
7	**3**	**100**	**300**	**0.949**	**1**
8	3	125	100	0.623	4
9	3	150	200	0.934	2

Bold indicates the sequnce of rank; Rank 1 indiactes the optimal ranking

3). The optimum responses for the Exp. No. 7 are MRR (57.274 g/min) and surface roughness (0.231 μm). The ideal combinations obtained using MOOSRA method gave maximum values of response characteristics for AWJM processes which have less influence on the performance of AWJM during JFRP composites. Further, ideal setting creates lesser ecological issues (because higher MRR and greater surface finish) as well as due to the use of green manufacturing process as AWJM which directly or indirectly results in the quality of the product, reduces the final product cost, and improves the machining performance of AWJM process. Hence, it is recommended that optimal setting, i.e., stand of distance as 3 mm, working pressure as 100 MPa, and nozzle speed as 300 mm/min, is used in the AWJM for effective machining of JFRP composites under green manufacturing scenario. Furthermore, the work suggested that the proposed method, i.e., MOOSRA, can be utilized for finding the best possible combination of input parameters for AWJM process in green manufacturing environment and other processes as well.

5.3.6 Confirmation Analysis

Further, to verify the obtained results via. MOOSRA method by using a confirmatory test is performed. The optimal setting, i.e., SOD (3 mm), working pressure (100 MPa), and nozzle speed (300 mm/min) are used for confirmatory experiments and the corresponding outcomes are shown in Table 5.7. The results show that confirmatory test results are comparable and acceptable with experimental outcomes for the ideal setting.

Additionally, machined surface morphology or surface integrity is done via scanning electron microscope (SEM). The surface integrity is done for the ideal case (i.e., Exp. No 7) proposed by MMOSRA method. The ideal condition taken as stand of

Table 5.7 Confirmatory tests for AWJM process of JFRP composites

Input conditions	Output parameters	Experimental results	Confirmatory tests results
Exp. No. 7 SOD (3 mm), WP (100 MPa) and NS (300 mm/min)	MRR (g/min)	57.274	57.012
	SR (μm)	0.231	0.229

distance as 3 mm, working pressure as 100 MPa and nozzle speed as 300 mm/min and SEM images of the same are shown in Fig. 5.6.

The SEM images show that machined surface at optimal condition found to be very good, smooth and uniform distribution of surface during the machining of AWJM [31, 34]. But a few places some surface imperfection like abrasive grains marks, abrasive impression marks, small voids, etc. are observed. These imperfections are mainly because of degree of interfacial adhesion between the matrix and fiber materials and also may be due to machining stress development in the composites. These issues can be minimized by homogenous mixing of fiber and matrix material, use of less minimal cutting pressure. Hence, the optimal setting gave acceptable surface roughness which justifies the results of the optimal conditions obtained via MOOSRA method.

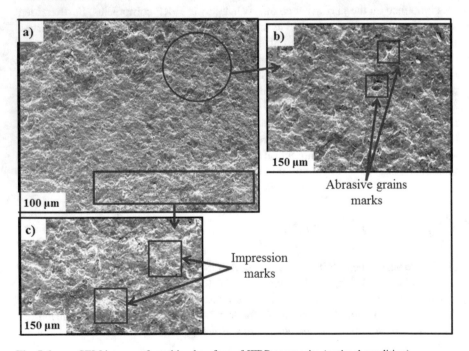

Fig. 5.6 a–c SEM images of machined surface of JFRP composite (optimal condition)

5.4 Summary

This chapter discussed the optimization and machinability characteristics of AWJM process factors on JFRP composites. Work considered MRR and surface roughness as response factors while processing of JFRP composites. Composite with 5% of fiber composition by weight with 5 mm composite thickness and specimen (workpiece) is prepared by the method of hand layup process and processing is done in AWJM process by changing the process parameters like SOD, WP and NS according to the Taguchi (L_9) orthogonal array method. The results show that process parameter SOD, working pressure, and nozzle speed are observed more significant parameters for AWJM process responses on JFRP composites. Optimal parameters obtained as stand of distance as 3 mm, working pressure as 100 MPa and nozzle speed as 300 mm/min which provides as MRR (57.274 g/min) and surface roughness (0.231 μm). Further, ANOVA and empirical models for MRR and surface roughness of JFRP composites are done and found that statistically significant and fit the data well enough with the experiment results with the 95% confidence level. At last, confirmatory test followed by surface morphology are performed for the optimal conditions results. It is observed that confirmatory results are similar and satisfactory with the present experimental results and also the machined surface found smooth and acceptable surface during the machining of AWJM.

From the above observations, it is concluded that the JFRP composite gives a better execution in the AWJM process. Ideal setting obtained via MOOSRA method can be used as standard setting and provides maximum values of response characteristics for AWJM processes which have less influence on the performance of AWJM during JFRP composites. Further, ideal setting creates lesser ecological issues due to the use of green manufacturing process as AWJM which directly or indirectly results in the quality of the product, reduces the final product cost, and improves the machining performance of AWJM process. Hence, it is recommended that optimal setting, i.e., stand of distance as 3 mm, working pressure as 100 MPa, and nozzle speed as 300 mm/min, is used in the AWJM for effective machining of JFRP composites under green manufacturing scenario. Furthermore, the work suggested that the proposed method, i.e., MOOSRA, can be utilized for finding the best possible combination of input parameters for AWJM process in green manufacturing environment and other processes as well. Moreover, the created empirical model for AWJM process responses can be implemented in an organized structural model for the prediction of MRR and surface roughness for all the cases.

References

1. K. Kumari, S.R. Devegowda, Trend analysis of area, production and productivity of jute in India. Phar. Innov. J. **7**(12), 58–62 (2018)
2. Y. Leonard, M.P. Ansell, Chemical Modification of Hemp, Sisal, Jute, and Kapok Fibers by Alkalization. J. App. Poly. Sci. **84**(12), 2222–2234 (2016)

3. H. Singh, J.I.P. Singh, S. Singh, V. Dhawan, S.K. Tiwari, A brief review of jute fibre and its composites. Mater. Today: Proc. **5**(14), 28427–28437 (2018)
4. V. Prasad, A. Joy, G. Venkatachalam, S. Narayanan, S. Rajakumar, Finite element analysis of jute and banana fibre reinforced hybrid polymer matrix composite and optimization of design parameters using ANOVA technique. Proc. Eng. **97**, 1116–1125 (2014)
5. P. Kaushik, Mittal K. Jaivir, Analysis of mechanical properties of jute fiber strengthened epoxy/polyester composites. Eng. Solid Mech. **5**(2), 103–112 (2017)
6. V. Chaudhary, P.K. Bajpai, S. Maheshwari, Studies on mechanical and morphological characterization of developed jute/hemp/flax reinforced hybrid composites for structural applications. J. Nat. Fibers **15**(1), 80–97 (2018)
7. T.M. Gowda, A.C.B. Naidu, R. Chhaya, Some mechanical propertiesof untreated jute fabric-reinforced polyester composites. Compos. Part A: Appl. Sci. Manuf. **30**(3), 277–284 (1999)
8. S. Luo, A.N. Netravali, Mechanical and thermal properties of environmentally friendly green composites made from pineapple leaf fibres and poly (hydroxybutyrate-co-valerate) resin. Poly. Comp. **20**(3), 367–378 (1999)
9. A.C. Karmaker, J.P. Schneider, Mechanical performance of short jute fiber reinforced polypropylene. J. Mater. Sci. Lett. **15**(3), 201–202 (1996)
10. T. Hojo, Z. Xu, Y. Yang, H. Hamada, Tensile properties of bamboo, jute and kenaf mat-reinforced composite. Energy Proc. **56**, 72–79 (2014)
11. P. Gondal, A. Verma, Y.I. Murthy, U. Mishra, Experimental study of sisal and jute fiber based biocomposite. Int. J. InterRese Innov. **6**(4), 398–404 (2018)
12. M.A. Ashraf, M. Zwawi, M.T. Mehran, R. Kanthasamy, A. Bahadar, Jute based bio and hybrid composites and their applications. J. Biocomop. **7**(9), 1–29 (2019)
13. M.H. Rezghi, M. Hamedi, M. Kubouchi, Y. Arao, Experimental study on drilling of jute fiber reinforced polymer composites. J. Compo. Mater. **53**(3), 283–295 (2019)
14. M. Ramesh, K. Palanikumar, K.H. Reddy, Experimental investigation and analysis of machining characteristics in drilling hybrid glass-sisal-jute fiber reinforced polymer composites, in *5th International & 26th All India Manufacturing Technology, Design and Research Conference (AIMTDR 2014), AIMTDR*, vol. 461(1), 461–466 (2014)
15. S. Kannappan, B. Dhurai, Investigating and optimizing the process variables related to the tensile properties of short jute fiber reinforced with polypropylene composite board. J. Eng. Fibers Fabrics **7**(4), 28–34 (2012)
16. R. Asekin, F. Tabassum, R. Shakif, M.S. Kaiser, S.R. Ahmed, Optimization of processing and post-processing conditions for improved properties of jute-fiber reinforced polymer composites. AIP Conf. Proc. **2121**, 140011-1–140011-17 (2019)
17. N. Johri, R. Mishra, H. Thakur, Design parameter optimization of Jute-chicken fiber reinforced polymeric hybrid composites. Mater. Today: Proc. **5**(9), 19862–19873 (2018)
18. Ray A. Jagadish, Optimization of process parameters of green electrical discharge machining using principal component analysis (PCA). Int. J. Adv. Manuf. Technol. **87**(5–8), 1299–1311 (2016)
19. S. Kalirasu, N. Rajini, S. Rajesh, J.T.W. Jappes, K. Karuppasamy, AWJM performance of jute/polyester composite using MOORA and analytical models. Mater. Manuf. Proc. **32**(15), 1730–1739 (2017)
20. J. Schwartzentruber, J.K. Spelt, M. Papini, Prediction of surface roughness in abrasive waterjet trimming of fiber reinforced polymer composites. Int. J. Mach. Tools Manuf. **122**, 1–17 (2017)
21. S. Chakraborty, A. Mitra, Parametric optimization of abrasive water-jet machining processes using grey wolf optimizer. Mater. Manuf. Proc. **33**(13), 1471–1482 (2018)
22. M. Naresh Babu, N. Muthukrishnan, Investigation on surface. Mater. Manuf. Proc. **29**(11–12), 1422–1428 (2014)
23. M. Mhamunkar, Optimization of process parameter of CNC abrasive water jet machine for titanium Ti 6Al 4 V material. Int. J. Adv. Res. Sci., Eng. Technol. **3**(3), 1640–1646 (2016)
24. T. Rajasekaran, K. Palanikumar, B.K. Vinayagam, Application of fuzzy logic for modeling surface roughness in turning CFRP composites using CBN tool. Prod. Eng. **5**(2), 191–199 (2011)

25. Bhowmik S. Jagadish, A. Ray, Prediction and optimization of process parameters of green composites in AWJM process using response surface methodology. Int. J. Adv. Manuf. Technol. **87**(5–8), 1359–1370 (2016)
26. R.V. Rao, V.D. Kalyankar, Optimization of modern machining processes using advanced optimization techniques: a review. Int. J. Adv. Manuf. Technol. **73**(5–8), 1159–1188 (2014)
27. M.K. Kulekci, U. Esme, O. Er, Y. Kazancoglu, Modeling and prediction of weld shear strength in friction stir spot welding using design of experiments and neural network. Materialwiss. Werkstofftech. **42**(11), 990–995 (2011)
28. W. König, C. Wulf, P. Grass, H. Willerscheid, Papers machining of fibre reinforced plastics. Ann. CIRP **34**(2), 537–548 (1985)
29. J. Kechagias, G. Petropoulos, N. Vaxevanidis, Application of Taguchi design for quality characterization of abrasive water jet machining of TRIP sheet steels. Int. J. Adv. Manuf. Technol. **62**(5–8), 635–643 (2012)
30. L.M.P. Durão, J.M.R.S. Tavares, V.H.C. Albuquerque, J.F.S. Marques, O.N.G. Andrade, Drilling damage in composite material. Materials **7**(5), 3802–3819 (2014)
31. M.A. Azmir, A.K. Ahsan, A. Rahmah, Effect of abrasive water jet machining parameters on aramid fibre reinforced plastics composite. Int. J. Mater. Form. **2**(1), 37–44 (2009)
32. Ray A. Jagadish, Green cutting fluid selection using Moosra method. Int. J. Res. Eng. Technol. **03**(15), 559–563 (2014)
33. S. Bhowmik, Jagadish, K. Gupta, Modeling and optimization of advanced manufacturing processes, in *Manufacturing and Surface Engineering*, 1st edn (Springer International Publishing, 2019). p. 1–74. https://doi.org/10.1007/978-3-030-00036-3
34. R. Kumar, K. Kumar, S. Bhowmik, Mechanical characterization and quantification of tensile, fracture and viscoelastic characteristics of wood filler reinforced epoxy composite. Wood Sci. Technol. **52**(3), 677–699 (2018)

Printed in the United States
By Bookmasters